# WORKBOOK FOR
# HUMAN FACTORS
# IN ENGINEERING
# AND DESIGN
## Third Edition

by

**MARK S. SANDERS**

California State University
Northridge, California

and

**ERNEST J. McCORMICK**

Purdue University
West Lafayette, Indiana

*to accompany*

**HUMAN FACTORS IN ENGINEERING AND DESIGN, Sixth Edition**

by Mark S. Sanders and Ernest J. McCormick
published by McGraw-Hill Book Company, New York, New York.

**KENDALL/HUNT PUBLISHING COMPANY**
2460 Kerper Boulevard   P.O. Box 539   Dubuque, Iowa 52004-0539

Our thanks to McGraw-Hill Book Company for the use of the same
cover design of their book, *Human Factors in Engineering and Design*,
Sixth Edition by Mark S. Sanders and Ernest J. McCormick.

Illustrations by Annette Angel.

ISBN 0-8403-4307-8

Printed in the United States of America
10   9   8   7   6   5   4   3   2

# Contents

# Preface

This *Workbook* is designed for use by students in courses which use the textbook, *Human Factors in Engineering and Design (6th Edition)* by Mark S. Sanders and Ernest J. McCormick (McGraw Hill Book Co., 1987). The *Workbook* is more than a study guide. The *Workbook* provides for each chapter in the text, key terms and concepts, review questions, activities, and one or more self contained projects.

The key terms/concepts and review questions are designed to complement one another. The review questions are broad summary and integration type questions.

The activities suggested for each chapter require resources beyond those that could be supplied in the *Workbook*. For the most part, they suggest things the student can do to experience some of the concepts and principles discussed in the text. Questions are asked to stimulate thinking about common experiences in terms of human factors principles.

The projects for each chapter are self contained units. No attempt was made to incorporate all the material in a chapter in a *Workbook* project, rather, a few salient portions of each chapter were selected around which the project was constructed. The projects can be roughly grouped into three types. The first, and most numerous, present materials to be analyzed, critique, and/or redesigned based on human factors concepts presented in the text. These projects make excellent material for class discussion.

Too often when students read a textbook well stocked with figures and tables, there is a tendency to just glance at the material and read on. The second type of *Workbook* project requires the student to actively use the figures and tables to solve specific "realistic" problems. This experience helps students discover and retain the basic relationships presented in the text, it gives them an opportunity to "explore" the figures and tables and uncover new insights into their meaning, and often it teaches them to appreciate the difficulty of using some of the figures to obtain accurate information.

The third type of *Workbook* project involves the carrying out of experiments related to material discussed in the text. Little or no apparatus is required to carry out the projects. The data can often be compared to results reported in the text. Experience has shown that it is one thing to read about the results of an experiment, but it is quite another to have been a subject and experienced directly the independent variables.

Each project is designed to enable the student to carry out the project on his/her own. In the case of all the projects, there is some references to relevant sections of the text. These sections will often be helpful, and usually be required in order for the student to complete the project in question.

The information to be furnished by the student for each project is to be entered in the spaces provided in the *Workbook*. Upon completion of the project, the sheet(s) are to be removed from the *Workbook* and turned into the instructor.

The *Workbook* contains a total of 508 terms/concepts, 98 review questions, 52 activities, and 28 projects. Comments and suggestions from users are encouraged and will be appreciated.

We would greatly appreciate receiving comments and suggestions for improving the *Workbook*, including suggestions for new projects and activities. Please send them to Mark S. Sanders, Psychology Department, California State University Northridge, Northridge, CA 91330.

# Chapter 1

# HUMAN FACTORS AND SYSTEMS

**Terms/Concepts:**

Human factors
Human engineering
Engineering psychology
Human Factors Society
System
Human-machine system
Manual, mechanical, and automated systems
System characteristics
System environment
System functions
System reliability
Mean time to failure
Series and parallel systems

**Review Questions:**

1. Discuss the focus, objectives, and approach of human factors.
2. Discuss the history of human factors.
3. Distinguish between, and give an example of a manual, mechanical, and automatic system.
4. What are the fundamental characteristics of a system?
5. What are the basic functions of a system and how do they relate to each other?
6. Distinguish between components in series and in parallel and discuss the implication for system reliability.

**Activities:**

1. Write to the Human Factors Society (P.O. Box 1369, Santa Monica, CA 90403) and ask for information on student memberships. Request a free copy of the Graduate Programs Directory if you are interested.
2. Go to the library and find recent issues of *Human Factors, Ergonomics,* and *Applied Ergonomics.* Look over the articles and assess what areas are generating research and development activities.
3. Describe a device or piece of equipment you own that could have been improved by better human factors design.

Project 1

## BASIC FUNCTIONS OF A SYSTEM

### Reading Assignment

Text: Chapter 1, pages 14–17.

### Purpose

The purpose of this project is to provide an opportunity to identify the basic functions of an actual operational system.

### Problem

A hospital can be considered a large complex system made up of many subsystems. One subsystem, which we will focus on, is the laboratory where various clinical tests (blood, urinalysis, etc.) are performed. This subsystem has all the characteristics of most systems.

The following is a short description of how a typical hospital laboratory functions. After reading the description, answer the questions on the following page.

---

Specimens arrive in the lab via four principal routes. Doctors will send patients to the lab with orders for a test or series of tests to be performed. If the test can be performed at that time, a laboratory assistant or technician will perform the test. If the test cannot be performed, an appointment is made for the patient to return and an information pamphlet is given to the patient concerning the tests he will undergo.

In some cases, "inpatient orders" are sent to the lab requesting a laboratory assistant to come to a patient's room in the hospital and obtain a specimen (blood, urine, etc.). In other cases, the specimen is taken by a nurse on the patient's floor and just the specimen is sent to the lab with an order form requesting various tests.

After a specimen arrives in the lab, a laboratory assistant reads the orders to determine which tests are to be performed. He may have to call requesting physicians because he cannot read their handwriting or to notify them if there will be an inordinate delay in obtaining the results of the tests. The assistant then obtains the needed supplies, e.g., slides, chemicals, test tubes, etc., and readies any required equipment. Some tests are very simple while others are complex and require multiple steps. The procedures are specified in department lab manuals. In many cases, the tests are automated. A specimen is simply placed in the test equipment and the results are read from various indicators on the machine. A time and materials card is filled out on each test and forwarded to Central Accounting for billing and inventory control. The results of each test are checked against a "norm chart" to determine if the test indicates any abnormalities. The results are transcribed onto lab report forms. One copy of the lab report is sent to the requesting physician, one copy to the Chart Room to be placed in the patient's hospital file, and one copy to the laboratory's own file.

This generally orderly process is often interrupted when a "stat" order is received. "Stats" are emergency orders which get the highest priority.

---

*Part A.* List some of the operational functions of the system.

_____

_____

_____

_____

*Part B.* Give several examples, within the operation of the laboratory, of each of the following.

1. Information receiving. _____

_____

_____

_____

2. Information storage. _____

_____

_____

_____

_____

3. Information processing and decision. _____

_____

_____

_____

_____

4. Action functions. _____

_____

_____

_____

5. Inputs and outputs between subsystems and components in the lab system and between the lab and

its environment. _____

_____

_____

_____

_____

# Chapter 2

# HUMAN FACTORS RESEARCH METHODOLOGIES

**Terms/Concepts:**

Descriptive studies
Criterion variables
Stratification variables
Representative sample
Random selection
Standard deviation
Correlation
Percentiles
Experimental research
Simulation
Independent variables
Dependent variables
Statistical significance

Evaluation research
Criterion measures
Terminal and intermediate criteria
Reliability of measurement
Face validity
Content validity
Construct validity
Contamination
Deficiency
Sensitivity
Human reliability
Data Store

**Review Questions:**

1. Distinguish between descriptive, experimental, and evaluation studies in terms of purpose, research setting, variables, and subjects.

2. List some criterion measures that could be used to evaluate a new home computer system. Include system-descriptive, task performance and human criteria.

3. Distinguish face validity, content validity, and construct validity. How do the concepts of contamination and deficiency relate to construct validity?

4. Discuss some human reliability data bases and the criticisms leveled at the measurement of human reliability.

**Activities:**

1. Go to the library and find one or two recent issues of *Human Factors, Ergonomics* and *Applied Ergonomics*. For each research article, determine: (1) whether it is a descriptive study, experimental research, or evaluation research; (2) whether it was conducted in a laboratory, field, or using simulation; (3) who the subjects were and how many were used; (4) what the variables were; (5) what type of criterion measures were used; and (6) what statistical analyses were conducted.

Project 2

## IDENTIFYING RESEARCH METHODOLOGIES

### Reading Assignment

Text: Chapter 2, pages 21–30.

### Purpose

The purpose of this project is to provide the student with an opportunity to classify research studies as descriptive, experimental, or evaluation research and to illustrate the range of studies in the science of human factors.

### Problem

The following is a list of study titles taken from *Human Factors, Ergonomics, Applied Ergonomics,* and the *Proceedings of the Human Factors Annual Meetings.* From the titles, classify each study as either descriptive (DESC), experimental (EXPER), or evaluation (EVAL) research. Write your choice in the space provided to the left of the item number.

_____ 1. An evaluation of adult clasping strength for restraining lap-held infants. (Mohan, D. & Schneider, L. *Human Factors,* 1979, 21, 635.)

_____ 2. Headlight glare resistance and driver age. (Pulling, N. & Wolf, E. *Human Factors,* 1980, 22, 103.)

_____ 3. Children's estimates of vehicle approach times. (Hoffmann, E. Payne, A., & Prescott, S. *Human Factors,* 1980, 22, 235.)

_____ 4. The positioning of type on maps: The effects of surrounding material on word recognition time. (Noyes, L. *Human Factors,* 1980, 22, 353.)

_____ 5. Effects of psychological load and speed on tractor operator error. (Sinden, J., Becker, W. & Shoup, W. *Applied Ergonomics,* 1985, 16, 183.)

_____ 6. Metabolic costs of stoopwalking and crawling. (Morrisey, S. George, C. & Ayoub, M. *Applied Ergonomics,* 1985, 16, 99.)

_____ 7. Office landscape: Does it work? (Brookes, M. *Applied Ergonomics,* 1972, 3, 224.)

_____ 8. A case study of the occupational stress implications of working with two different actuation/safety devices. (Salvendy, G. Shodja, S., Sharit, J. & Etherton, J. *Applied Ergonomics,* 1983, 14, 291.)

_____ 9. Effect of intonation form and pause durations of automatic telephone number announcements on subjective preference and memory performance. (Watherworth, J. *Applied Ergonomics,* 1983, 14, 39.)

_____ 10. The effect of display format on the direct entry of numerical information by pointing. (Long, J., Whitefield, A. & Dennett, J. *Human Factors,* 1984, 26, 3.)

_____ 11. Comparison of abbreviation methods: measures of preference and decoding performance. (Rogers, W. & Moeller, G. *Human Factors,* 1984, 26, 49.)

_____ 12. Effects of adjustable furniture on VDT users. (Shute, S. & Starr, S. *Human Factors,* 1984, 26, 157.)

_____ 13. Short-term changes in eyestrain of VDU users as a function of age. (Hedman, L. & Briem, V. *Human Factors*, 1984, 26, 357.)

_____ 14. Synthesized speech rate and pitch effects on intelligibility of warning messages for pilots. (Simpson, C. & Marchionda-Frost, K. *Human Factors*, 1984, 26, 509.)

_____ 15. Television viewing at home: Distances and visual angles of children and adults. (Nathan, J., Anderson, D., Janaro, R. & Bechtold, S. *Human Factors*, 1985, 27, 467.)

_____ 16. The effect of vibration frequency and direction on the location of areas of discomfort caused by whole-body vibration. (Whitham, E. & Griffin, M. *Applied Ergonomics*, 1978, 9, 231.)

_____ 17. A laboratory study of the effects of moderate thermal stress on the performance of factory workers. (Meese, G., Kok, R., Lewis, M. & Wyon, D. *Ergonomics*, 1984, 27, 19.)

_____ 18. Do TV pictures help people to remember the weather forecast? (Wagenaar, W., Schreuder, R. & von der Heijden, A. *Ergonomics*, 1985, 28, 765.)

_____ 19. Effects of local vibration transmitted from ultrasonic devices on virbotactile perception in the hands of therapists. (Lundstrom, R. *Ergonomics*, 1985, 28, 793.)

_____ 20. Driver behaviour in the presence of child and adult pedestrians. (Thompson, S., Fraser, E. & Howarth, C. *Ergonomics*, 1985, 28, 1469.)

_____ 21. How people create spreadsheets. (Brown, P. & Gould, J. *Proceedings of the Human Factors Society 30th Annual Meeting*, 1986, 1, 29.)

_____ 22. Effects of benign experiences on the perception of risk. (Karnes, E., Leonard, S. & Rachwal, G. *Proceedings of the Human Factors Society 30th Annual Meeting*, 1986, 1, 121.)

_____ 23. The influence of position, highlighting, and imbedding on warning effectiveness. (Strawbridge, J. *Proceedings of the Human Factors Society 30th Annual Meeting*, 1986, 1, 716.)

_____ 24. Human factors engineering analysis of Marine Corps night attack aircraft. (Breitmaier, W. & Waldrop, G. *Proceedings of the Human Factors Society 30th Annual Meeting*, 1986, 1, 861.)

_____ 25. Torque capabilities of the elderly in opening screw-top containers. (Imrhan, S. & Loo, C. *Proceedings of the Human Factors Society 30th Annual Meeting*, 1986, 1, 1167.)

_____ 26. Human factors assessment: M9 armored combat earthmover (ACE). Krohn, G. *Proceedings of the Human Factors Society 30th Annual Meeting*, 1986, 1, 1311.)

_____ 27. On difficulties in localizing ambulance sirens. (Caelli, T. & Porter, D. *Human Factors*, 1980, 22, 719.)

_____ 28. Comparisons of three field methods for measuring oxygen consumption. (Louhevaara, V., Ilmarinen, J. & Oja, P. *Ergonomics*, 1985, 28, 463.)

_____ 29. Air traffic control using a microwave landing system. (Gershzohn, G. *Human Factors*, 1980, 22, 621.)

_____ 30. Visual effects of wall colours in living rooms. (Kunishima, M. & Yanase, T. *Ergonomics*, 1985, 28, 869.)

# Chapter 3

# INFORMATION INPUT AND PROCESSING

**Terms/Concepts:**

Information
Bit
Redundancy
Bandwidth
Hick-Hyman law
Distal stimuli
Proximal stimuli
Direct sensing
Indirect sensing
Coded stimuli
Reproduced stimuli
Display
Static information
Dynamic information
Quantitative information
Qualitative information
Status information
Representational information
Identification information
Time-phased information
Stimulus dimension
Absolute judgments
Relative judgments

Magical number $7 \pm 2$
Orthogonal dimensions
Redundant dimensions
Detectability
Discriminability
Conceptual compatibility
Movement compatibility
Spatial compatibility
Modality compatibility
Signal Detection Theory
Response criterion
Beta
d'
Sensory storage
Working memory
Chunk
Long-term memory
Selective attention
Focused attention
Divided attention
Single versus multiple resource models
Mental workload

**Review Questions:**

1. How is the concept of information defined within information theory?
2. What are the major classifications of information presented by displays?
3. How can stimulus dimensions be combined to enhance absolute judgments and what is the effect of combining dimensions on the number of stimuli that can be discriminated on an absolute basis?
4. What are the characteristics of a good coding system?
5. What is compatibility, what are its origins, and how do we identify compatible relationships?
6. How does signal detection theory explain false alarms, hits, misses, and correct rejections in a target detection task?
7. Distinguish between selective attention, focused attention, and divided attention; and indicate some methods for improving performance in each type of task.
8. What are the four major categories under which measures of mental workload are classified?

**Activities:**

1. Try to read a book and at the same time *listen* to a news broadcast on the radio. Did you read slower when you were listening to the radio? Now try to read while you try to *ignore* a news broadcast on the radio. Did you read slower when you were ignoring the radio or when you were trying to listen to it? How do these situations relate to focused and divided attention?

Project 3

# THE EFFECT OF STIMULUS SET SIZE ON ABSOLUTE DISCRIMINATIONS

## Reading Assignment

Text: Chapter 3, pages 50–53; 57–59.

## Purpose

The purpose of this project is to demonstrate the effect of set size on the ability to make absolute judgments of size.

## Problem

This project is an experiment to test your ability to make absolute judgments of size. The task is repeated four times, each time increasing the number of different sized stimuli which must be judged.

*Part A.* Determine the effect of set size on ability to make absolute judgments.

Procedure:

1. Cut out the numbered strip guide at the top of Figure 1. Place this on a table face up. This will serve as a guide for classifying the stimuli.
2. Cut out the eighteen 1 inch squares in columns 1, 2, and 3 of Figure 1.
3. Mix the squares together and place them face down on the table.
4. Pick up one square at a time and decide, based on the size of the circle, which column it came from (1, 2, or 3) and place it FACE DOWN below the appropriate number on the strip guide.
5. Do this for each of the eighteen items.
6. When finished, turn over the piles and count the number of errors, i.e., the number of items placed in the wrong pile. (Notice, it is much easier to determine errors when you can see the other items. This, of course, is because relative judgments are much easier to make than are absolute judgments.) Record the number of errors in Table 1 for a stimulus set size of 3.
7. Now cut out the squares in columns 4 and 5 of Figure 1 and mix them together with the squares you already cut. Place them all face down on the table.
8. Repeat step 4, placing the items face down below the appropriate columns 1, 2, 3, 4, or 5.
9. Determine the number of errors and record in Table 1 for set size of 5.
10. Cut out columns 6 and 7 from Figure 1, mix them together with the items from columns 1 to 5 and place them all face down on the table.
11. Sort them into the appropriate piles a you did in steps 4 and 8. Score the number of errors and record it in Table 1 for a set size of 7.
12. Finally, cut out columns 8 and 9 from what is left of Figure 1, mix them with the previously used items, place them all face down, and sort them one at a time as you did before into their appropriate piles. Determine the number of errors and record in Table 1.
13. Determine the *percentage* of errors made with each set by dividing the number of errors, column 2 in Table 1, by the number of possible errors listed in column 3 and multiply by 100. Record in column 4 of Table 1.

## TABLE 1.
### Absolute Discrimination Data Sheet

| (1)<br>Set Size | (2)<br>Number of errors | (3)<br>Total Possible Errors | (4)<br>Percentage [(2) ÷ (3)] × 100 |
|---|---|---|---|
| 3 | _____ | 18 | _____ |
| 5 | _____ | 30 | _____ |
| 7 | _____ | 42 | _____ |
| 9 | _____ | 54 | _____ |

*Part B*. Discuss your results in terms of the "magical number seven, plus or minus "two" presented in the text on page 51

_____

_____

_____

_____

_____

_____

_____

*Part C*. Where did you tend to make most of the errors with the 9 item stimulus set? With the small sized circles? Large sized circles? Or with the middle range sizes? Why do you think this happened?

_____

_____

_____

_____

_____

_____

_____

_____

Name _____ Course _____ Seat Number _____ Date _____

| SMALLEST | | | | | | LARGEST |
|---|---|---|---|---|---|---|---|---|
| 1 | 2 | 3 | 4 | 5 | 6 | 7 | 8 | 9 |

Column number

| 1 | 2 | 3 | 4 | 5 | 6 | 7 | 8 | 9 |
|---|---|---|---|---|---|---|---|---|

**Figure 1.** Cut out strip guide and squares according to instructions.

# Chapter 4
# VISUAL DISPLAYS OF STATIC INFORMATION

**Terms/Concepts:**

Visual acuity
Accommodation
Minimum separable acuity
Visual angle
Vernier acuity
Minimum perceptible acuity
Stereoscopic acuity
Convergence
Dark adaptation
Luminance contrast
Dynamic visual acuity
Visibility

Legibility
Readability
Stroke width
Irradiation
Width-height ratio
Segmented character displays
Dot matrix displays
Scan lines
Pixels
Density of alphanumeric material
Coding dimensions
Confusion matrix

**Review Questions:**

1. Diagram the eye and identify the principle parts.
2. What factors influence visual acuity and visual discriminations?
3. What are some of the typographical factors that affect the readability of alphanumeric displays?
4. Discuss the use of color in coding and indicate factors that mediate the effects of color coding on performance.
5. What are three methods for evaluating symbolic signs?

**Activities:**

1. Cut out newspaper headlines, subheads, and print of various sizes. Have someone stand against a wall and try to read one size of type as you hold it up at a distance. Slowly move closer and closer until the person can read it almost without error. Measure the viewing distance from the person's eyes to the copy and the letter height. Do this little experiment with each size of type you found. How do the results compare to those given in Figure 4-12 on page 93 of the text?
2. Look at various makes and models of automobiles (especially foreign makes) and observe the symbols used for various controls and displays such as windshield wipers and washers, high and low beam headlights, defrosters, air conditioning controls, etc. How consistent are the symbols between makes and models? Which are hard to understand? Which are easy to comprehend?
3. Go to a computer store and look at the various makes and models of computer screens. Try to read text from the various models. Which screens were easier to read text from? Why? Try to determine the size of the dot matrix making up the letters of the various screens and the overall size of the letters. Which matrix size was easier to read? Did the size of the letters make much of a difference in ease of reading?

Project 4-A

## EFFECT OF CLOSE-SET TEXT ON READING SPEED

### Reading Assignment

Text: Chapter 4, pages 85–98.

### Purpose

The purpose of this project is to replicate an experiment performed by Moriarty and Scheiner (1984) and discussed on page 94 of the text. The purpose of the replication is to demonstrate the effects on reading speed of varying the space between letters in text copy.

### Problem

Figures 1 and 2 are excerpts of text printed with different letterspacings. Figure 1 is set in normal letterspacing and Figure 2 is set in close-set text. Both figures are set in 11-point Helvetica font. This project will replicate, in part, the Moriarty and Scheiner experiment. You are to perform the experiment using two subjects (two friends, or one friend and yourself). You will need a stop watch or a watch that indicates seconds.

Instructions:

1. Instruct the first subject that when you say "start" to read, not skim, the text in Figure 1 at a comfortable rate and to mark with a slash the last word they read when you tell them to stop.
2. Say "start" and time the subject for 1 min 30 sec and say "stop."
3. Repeat steps 1 and 2 using the same subject but have the person read the text in Figure 2.
4. Determine the number of words read for Figure 1 and Figure 2 and record them in Table 1. The number at the end of each line in Figures 1 and 2 is the cumulative number of words from the beginning of the text to the end of that line.
5. Repeat steps 1 through 4 with a second subject, except reverse the order in which the person reads the texts. That is, have the person read Figure 2 first and then Figure 1. This is a form of counterbalancing and helps control for practice and fatigue effects.
6. Compute the mean number of words read in each condition and record them in Table 1.

**TABLE 1**
Data from Experiment (Number of Words Read in 1.5 Min)

| | Letterspacing | |
| Subject | Normal (Figure 1) | Close-set (Figure 2) |
|---|---|---|
| 1 | | |
| 2 | | |
| Mean | | |

*Part A:* Which condition resulted in more words being read? Was this what was found by Moriarty and Scheiner? Why did you think you got the results you did?

_____

_____

_____

_____

_____

_____

_____

**Figure 1 Normal Letterspacing**

We said earlier that machine-environment 6 problems should be examined from a 12 systems viewpoint, and that activities are a 19 major focal point of our interest. The 26 performance of activities is influenced by 32 the interaction of many behavioral and 38 design factors. While a start has been 45 made by human factors researchers who 51 have dealt with some machine-environment 57 problems machine-environment researchers 61 face a formidable challenge in their 67 attempts to better understand the nature of 74 these relationships. 76

Human factors, like architecture, includes 81 problem areas ranging from the "macro" to 88 the "micro" scale. At the macro scale, the 96 concern is how human performance relates 102 to an overall system. For example, a 109 problem faced by both equipment and 115 building designers is the need to integrate 122 manually operated controls for best system 128 performance. For example, effective lighting 133 depends upon the ability to control the 140 quantity and quality of light where it is 148 needed—e.g., to perform a task. A further 156 desirable characteristic, in view of the 162 concern for energy conservation, is to have 169 lighting in areas only where it is required. 177 Consequently, area control of lighting is 183 more desirable than a single control for a 191 partially occupied large area. The architect 197 must then determine where to place light 204 switches (or other controls) for maximum 210 effectiveness. 211

At the micro scale, a major concern is 219 the proper design of products (e.g., 225 machines, furniture). This interest is 230 synonymous with that of the industrial 236 designer. Between the macro and micro 242 scales, researchers have addressed 246 problems of permanent and temporary 251 features of buildings, such as determining 257 the proper dimensions for shelves and 263 kitchen work surfaces. The common feature 269 in such investigations is the focus on 276 performing a given activity. 280

F. Taylor (machinist and plant supervisor) 286 is often termed the father of scientific 293 management. He formulated the idea of 299 conducting analytic studies which were 304 designed to increase the productivity of 310 workers. His approach was to standardize 316 work procedures, based on the way that 323 the best workers performed their jobs. The 330 work activities and rest pauses were 336 specified completely, with the worker 341 treated virtually as a machine. 346

Time and motion studies were used in the 354 1920's by F. Gilbreth and his wife, L. 362 Gilbreth to break down jobs into their 369 elementary operations. The basic concept 374 was that work productivity differed greatly 380 from person to person because of individual 387 differences in skills. Differences in work 393 methods may be best analyzed by slow 400 motion photography, both for finding the 406 errors of less skilled workers and for 413 teaching to new employees the superior 419 methods of the most highly skilled workers. 426 There are general principles and 431 procedures that may be applied to all jobs. 439 At first, they employed a stopwatch and 446 described the nature of the motions used. 453 Later, they recorded minute movement 458 using photographic procedures. Movements 462 were charted into 16 elements of motion, 469 said to be the basic elements of industrial 477 processes. 478

The general approach developed by the 484 Gilbreths was later used by human factors 491 researchers to analyze activities. The 496 procedure is called task analysis. It is used 504 to identify the factors necessary to 510 adequately complete a task. The goal is to 518 define the critical activities occurring in a 525 work situation in such a way as to provide a 535 sound basis for performance evaluation. 540 Another purpose is to aid in modifying job 548 operations in order to facilitate performance 554 and avoid error. 557

**Figure 2. Close-set Letterspacing**

Many work and other activities require 6
people to obtain information from their 12
environment. Depending on the nature of the 19
information obtained, they select a course of 26
action from several alternatives. In human 32
factors research, the device used to convey 39
information is termed a *display.* The most 46
commonly used displays depend on visual 52
and/or auditory signals. Such displays are 59
commonly employed in cooking tasks: a light 66
which identifies the stove burner which is on; a 75
buzzer which is preset on an oven to indicate 84
that the required baking time has elapsed. 91

While a display indicates that an action 98
should be taken, *controls* serve the purpose of 106
enabling the person to take action. Control 113
design has been of major human factors 120
interest because of the variety of movements 127
and forces that can be applied by users. 135
Speed and accuracy data have been collected 142
on tracking tasks (gunnery, radar) where the 149
motion of the operator is continuous. 155
Anthropometric measures of arm, leg, and 161
back strength have been compiled in the 168
design of levers, and automobile emergency 174
brakes, as well as foot brakes. Other 181
measures which have been made include grip, 188
elbow, and shoulder strength. Relationships 193
between direction of motion and strength, while 200
seated, standing and lying down, have also 207
been systematically compiled. 210

A systems approach is employed in the 217
design of a configuration of controls and 224
displays. There are several possible principles 230
of arrangement. In a functional arrangement, 236
the grouping is in accordance with the activity 244
being performed. (In an aircraft, all of the flight 253
instruments can be placed together.) Another 259
grouping might be on the basis of importance, 267
where critical displays and controls would be 274
placed in the most easily accessible positions. 281
The best location is another possibility with 288
"best" depending on convenience, accuracy, 293
speed, or strength. A sequence-of-use method 301
is based on patterns of normal operation 308
(items physically positioned in accordance with 314
sequence of use). Finally, a frequency-of-use 322
principle can be used. 326

The determination of the proper relationship 332
between manual controls and displays has 338
received considerable attention. The activity is 344
one of simultaneous monitoring and control. An 351
operator is typically required to monitor several 358
displays and make control adjustments as the 365
circumstances demand. Jobs of this type, 371
which were formerly associated with military, 377
space, and aircraft operations, are now often 384
found in high rise buildings with centralized 391
communications, fire and surveillance systems. 396
Moreover, the problem is one which is 403
constantly addressed by those designing 408
appliances and other apparatus used in 414
buildings. 415

Consider a classroom. The experimenter is 421
concerned with identifying critical 425
environmental characteristics which might 429
affect the performance of the teachers and/or 437
the students. As a passive observer, the 444
researcher would record data on natural 450
behaviors with minimal interference (e.g., from 456
behind a one-way mirror in a wall of a room). 467
These behaviors are likely to be complex and 475
hard to define. For example, the teacher could 483
be observed "speaking to the class," "moving 490
about the room" and the students could be 498
"attentive" or "restless." In most instances, a 505
limited number of subjects or activities is 512
watched for relatively long periods of time. 519
Since many events occur during such a 526
situation, it is often difficult to obtain agreement 534
among researchers as to what should be 541
observed, or what records should be made of 549
observations. Standardized procedures are 553
developed to overcome this, together with 559
careful selection and training of observers. 565
Observations are generally made of public, 571
natural and therefore highly visible events 577
rather than private ones. After making a 584
number of observations, particular activities 589
and environmental characteristics might be 594
examined, e.g., the effects of glare produced 601
by sunlight on reading activities. 606

Project 4-B

## SHAPE CODING EXPERIMENT

### Reading Assignment

Text: Chapter 4, pages 98–103.

### Purpose

The purpose of this project is to familiarize the student with a typical experimental procedure used in coding research. Further, the student may come to appreciate more the importance of selecting code symbols for displays which are easily distinguished from one another.

### Problem

The text on pages 99–100 discusses the results of a coding experiment carried out by Smith and Thomas (1964) and presents the results in Figure 4-21 (page 100). This project involves a partial "replication" of that experiment and a comparison of the results with those obtained in the Smith and Thomas study. You can use yourself as a subject in this experiment, or someone else.

Instructions:

1. Each of the three figures in this project, Figures 1, 2, and 3, is made up of 100 code symbols. Each figure is composed of one class of symbols shown in Figure 4-20 (page 99 of the text). Figure 1 is composed of "military symbols," Figure 2 of "geometric forms," and Figure 3 of "aircraft shapes."

2. The task is to count the number of a specific type of symbol in each figure. The "target symbol," the one to be counted, is given in Table 1. A total of four separate counts, using different "target symbols," is to be made on each of the three figures. Do not make all four counts on the same figure consecutively. Count a symbol from Figure 1, then Figure 2, etc.

3. For each count, work as quickly and accurately as possible. Use a watch with a second hand and time yourself, or your subject, on each count. Do not make any marks on the figures while counting. Hold the figures so that the wire binding of the *Workbook* is at the top when counting.

4. For each count, record in Table 1 the number of seconds it took to make the count, and the actual count itself.

5. At the bottom of page 22 (inverted) is the correct count for each target symbol used. (Don't peek now. Wait until you have completed the experiment.) Compare your count with the correct count and record in Table 1 whether that trial was correct (C) or an error (E).

6. Compute the mean time to count the four target symbols for each of the three figures. Record these in Table 2.

7. Determine the percentage of trials which were in error (number of trials in error ÷ 4) for each of the three figures. Record these in Table 2.

8. Compare your data to the data given in Figure 4-21 on page 100 of the text. Use the "X's" at density 100 to make the comparisons. The "X's" indicate data from displays in which all code symbols were the same color which is closest to our all black condition.

9. Answer the following questions.

What similarities and differences existed between your data and that presented in the text?

_____

_____

_____

Which *class* of symbols (i.e., aircraft, geometric, or military) did you find most difficult to count? Why was it so difficult?

_____

_____

For the military symbols, which was the most difficult to count? Why was it so difficult?

_____

_____

What effect do you think a *redundant* color code would have on time and errors for this task?

_____

Aircraft shapes: 1(17); 2(26); 3(16); 4(20).
Geometric forms: 1(26); 2(15); 3(20); 4(19).
Military symbols: 1(24); 2(19); 3(16); 4(19).

22

## TABLE 1
Worksheet to Record Data from Experiment

| Target Symbol to be Counted | Time (sec) | Your Actual Count | Correct (C) or Error (E) |
|---|---|---|---|
| **Figure 1. Military symbols:** | | | |
| 1. Gun | _____ | _____ | _____ |
| 2. Missile | _____ | _____ | _____ |
| 3. Radar | _____ | _____ | _____ |
| 4. Ship | _____ | _____ | _____ |
| **Figure 2. Geometric forms:** | | | |
| 1. Triangle | _____ | _____ | _____ |
| 2. Semicircle | _____ | _____ | _____ |
| 3. Star | _____ | _____ | _____ |
| 4. Diamond | _____ | _____ | _____ |
| **Figure 3. Aircraft shapes:** | | | |
| 1. F-100 | _____ | _____ | _____ |
| 2. C-54 | _____ | _____ | _____ |
| 3. B-52 | _____ | _____ | _____ |
| 4. C-47 | _____ | _____ | _____ |

## TABLE 2
Summary of Experimental Data

| | Average Time | Percent of Trials in Error |
|---|---|---|
| Military symbols | _____ | _____ |
| Geometric forms | _____ | _____ |
| Aircraft shapes | _____ | _____ |

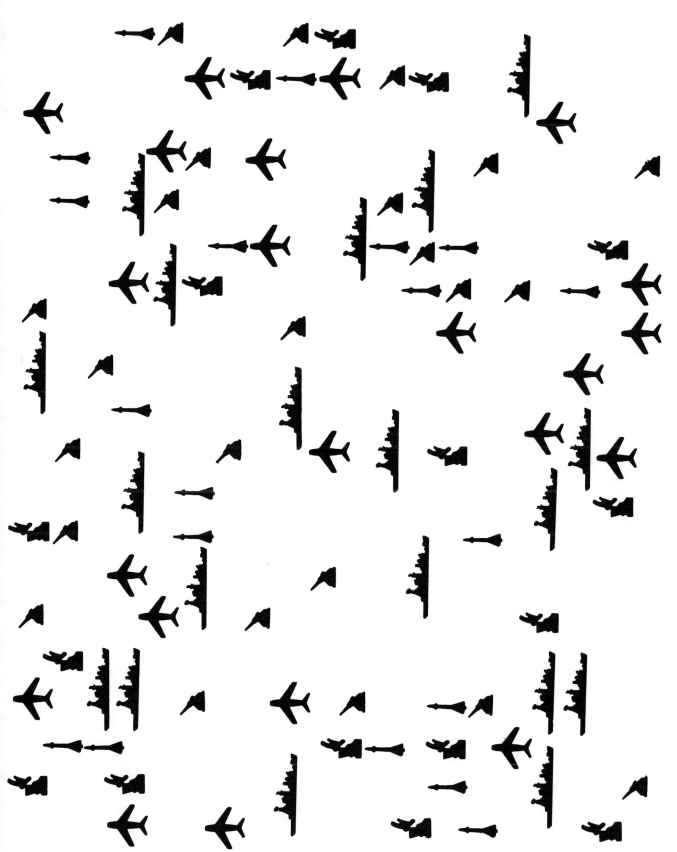

**Figure 1.** Military Symbols.

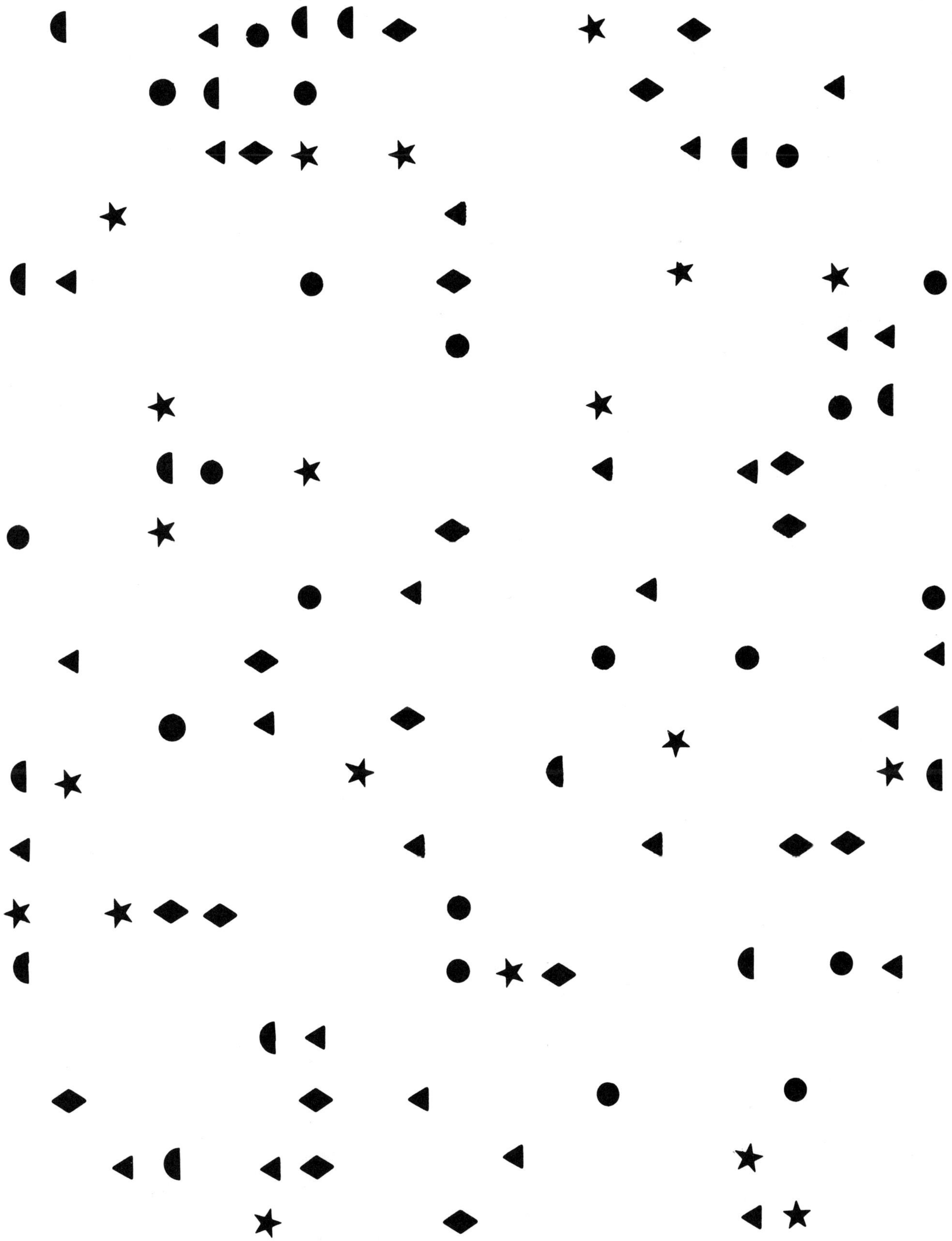

**Figure 2.** Geometric Forms.

**Figure 3.** Aircraft Shapes.

Project 4-C

## SYMBOLIC SIGNS

Reading Assignment

Text: Chapter 4, pages 103–110.

Purpose

The purpose of this project is to demonstrate the need to make signs suggestive of what they are intended to symbolize. Actual signs, used in various countries, are presented and the student is asked to indicate their meaning.

Problem

The following signs either have been proposed, have actually been used, or are currently used in transportation related facilities (airports, train stations, etc.) in various countries.[1] Some of the examples were chosen to illustrate poor design.

Beside each sign, indicate what you think it symbolizes. Several of the signs mean the same thing. Don't change your answers, but put down your first impression. The letters under each sign indicate the source of the sign. The code is as follows:

AIGA: American Institute of Graphic Arts recommendation.[1]

BAA: British Airport Authority (Heathrow Airport).

D/FW: Dallas-Fort Worth International Airport.

IATA: International Air Transport Association.

NRR: Netherlands' Railroads

O'64: Tokyo Olympic Games.

O'72: Munich Olympic Games (and Frankfort International Airport).

PORT: Port Authority of New York and New Jersey (JFK Airport).

TA: Tokyo Airport.

TC: Transport Canada (Airport facilities).

UIC: International Union of Railways (European railway facilities).

WO'72: Sapporo Winter Olympics (Japan).

Y'67: Montreal Expo.

The correct answers are shown on page 34 the last page of this project.

1. The American Institute of Graphic Arts. *Symbol Signs: The development of passenger/pedestrian oriented symbols for use in transportation-related facilities.* DOT-OS-40192. Springfield, Va: National Technical Information Service, November 1974.

1. _____

X'67

AIGA

2. _____

3. _____

TC

IATA

4. _____

5. _____

AIGA

D/FW

6. _____

7. _____

UIC

WO'72

8. _____

9. _____

IATA

TA

10. _____

11. _____

AIGA

12. _____

NRR

13. _____

O'72

14. _____

O'64

15. _____

O'72

16. _____

O'72

17. _____

O'64

18. _____

O'72

19. _____

PORT AIGA

20. _____

BAA

1.  No smoking.

2.  Rail transportation.

3.  Do not enter.

4.  Elevator.

5.  Hotel information.

6.  Gift shop

7.  Bar.

8.  Money exchange.

9.  Lost and found.

10. Men's restroom.

11. Customs.

12. Do not enter.

13. Baggage check-in.

14. Information.

15. Elevator.

16. No parking.

17. Lockers.

18. Hotel information.

19. Rental car.

20. Restrooms.

# Chapter 5

# VISUAL DISPLAYS OF DYNAMIC INFORMATION

**Terms/Concepts:**

Scale unit
Fixed scale, moving pointer displays
Moving scale, fixed pointer displays
Digital displays
Electronic displays
Qualitative scales
Check reading

Just noticeable difference (JND)
Representational displays
Principle of pictorial realism
Principle of integration
Principle of compatible motion
Principle of pursuit presentation

**Review Questions:**

1. Discuss the advantges and disadvantages of the various types of quantitative displays.

2. What specific features of conventional quantitative displays influence performance of people using them?

3. Discuss some factors that affect the detectability of warning lights.

**Activities:**

1. Find examples of good and not-so-good visual displays used to present dynamic information. Critique them from a human factors perspective. Some places to look include automobiles, electronic test equipment, and industrial control panels.

Name _____ Course _____ Seat Number _____ Date _____

Project 5

## DESIGN OF VISUAL DISPLAYS

Reading Assignment

Text: Chapter 5, pages 117–124.

Purpose

The purpose of this project is to afford the student an opportunity to apply information from the text to the design of typical visual displays.

Problem 1

Figure 1 represents a tachometer used in testing rotary equipment. The operational range is 50 rpm. The operator must read the scale to the nearest .5 rpm (the scale unit is .5 rpm).

*Part A*. Critically evaluate the instrument dial in Figure 1 on the following points, indicating poor design practices if present.

1. Number of scale markers. _____

_____

2. Numerical progression. _____

_____

3. Design of the pointer. _____

_____

4. Other aspects of the dial. _____

_____

*Part B*. Redesign the dial in the space provided at the right in Figure 1 following recommended design practices. Put in all major and/or numbered markers. Minor markers need be illustrated only between the first two major or numbered markers.

**Figure 1.** Tachometer used in testing rotary equipment. The scale unit is .5 rpm. The operational range is 50 rpm.

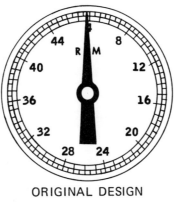

ORIGINAL DESIGN          NEW DESIGN

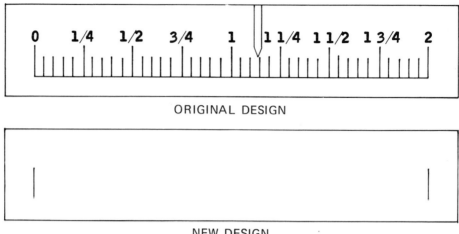

**Figure 2.** Steam gauge used to measure pressure in a boiler. The scale is 1/20th of a pound. The maximum pressure possible is two pounds.

## Problem 2

Figure 2 represents a steam gauge used to measure pressure in a boiler. The maximum pressure possible is two pounds. The scale must be read to the nearest 1/20th or .05 of a pound.

*Part A.* Critically evaluate the instrument dial in Figure 2 on the following points, indicating poor design practices if present.

1. Number of scale markers. _____

_____

2. Numerical progression. _____

_____

3. Design of the pointer. _____

_____

4. Other aspects of the dial. _____

_____

*Part B.* Redesign the dial in the space provided in Figure 2 following recommended design practices. Put in all major and/or numbered markers.

## Problem 3

Illustrate how the *quantitative* instrument illustrated below in Figure 3 can be converted to a *qualitative* instrument by modifying its dial face. The desired interpretation of the quantitative readings are indicated in the space adjacent to the illustration.

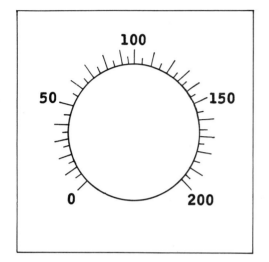

| Interpretation | Range of Readings |
|---|---|
| Normal | 0 to 70 |
| Caution | 70 to 120 |
| Danger | 120 to 200 |

**Figure 3.** Quantitative instrument to be converted to a qualitative instrument. (For this purpose, show the changes on the instrument itself.)

# Chapter 6

# AUDITORY, TACTUAL, AND OLFACTORY DISPLAYS

**Terms/Concepts:**

Frequency
Sinusoidal (sine) wave
Hertz
Pitch
Intensity
Loudness
Decibel
Sound pressure level
Signal-to-noise ratio
Sound spectrum
Bandwidth
Pinna
Meatus
Tympanic membrane
Malleus, incus, and stapes
Oval window

Tensor tympani muscle
Stapedius muscle
Acoustic or aural reflex
Cochlea
Basilar membrane
Organ of Corti
Auditory nerve
Masking
Stereophony
Pathsounder, Single-Object Sensor, and
Sonic Pathfinder
Sonicguide
Two-point threshold
Optacon
Olfactory epithelium

**Review Questions:**

1. Discuss the types of information best presented through auditory, tactual, and olfactory displays. What are the inherent advantages and limitations of these types of displays?
2. Diagram the ear and trace the path of a sound from the pinna to the brain.
3. What are some guidelines for designing auditory signals for detection and relative discrimination?
4. List some specific applications for tactual and olfactory displays.

**Activities:**

1. Next time you are riding in a car and you hear a siren, try to localize the direction from which the siren is coming. Was it difficult to do? Did you move your head from side to side as an aid in localizing the sound?
2. Perform the little stereo-speaker demonstration described on page 153 of the text.
3. Using two nails or pins, measure the two-point threshold on various parts of the hand and forearm of a friend. Compare the results with those shown in Figure 6-12, page 161 of the text.
4. Release an odorous substance (e.g., perfume, ammonia, banana oil) in a room and determine how long it takes for a person to detect the odor across the room. How long does it take before you adapted to the smell?

Project 6

## SPATIAL ORIENTATION FROM A TACTUAL MAP

### Reading Assignment

Text: Chapter 6, pages 164–166.

### Purpose

The purpose of this project is to demonstrate how tactile maps can be used to convey spatial information and to illustrate the difficulty in conveying such information with a tactual display.

### Problem

In this project you are asked to construct a simple tactile map of a hypothetical campus. Using this map, you are to test one or more people on their ability to draw the map after being given an opportunity only to feel it.

Procedure:

1. Cut out the shapes in Figure 1, the roadway and buildings, and carefully trace them or paste them on a thin piece of cardboard (the back cover of this workbook can be used).
2. Now carefully cut the shapes out of the cardboard and neatly paste them where indicated on Figure 2. With the shapes pasted on Figure 2, it becomes a somewhat crude, but effective, tactile map of Tactile U.
3. Have a friend serve as a subject. *Do not let the person see the map.* Blindfold the person and read the following instructions:

   I will present you with a tactile map of a hypothetical university. The buildings and major walkways are cardboard pasted on the paper. Your task is to feel the map with your fingers and try to get a mental picture of it. Use both hands if you wish. Pay attention to the relative positions of the buildings and walkways, their shape and the distance between them. You will have 5 minutes to study the map with your fingers. When I say stop, I will remove your blindfold and give you a blank piece of paper and a pencil. You are to draw the map as accurately as you can. Note, there are seven buildings on the map. Are there any questions?

4. Place Figure 2 in front of the subject with the "Name . . . Course . . ." at the top of the page. Place their hands on the map. Say "start" and time them for 5 minutes, then say "stop." Remove Figure 2 from view before you remove their blindfold.
5. Give the person a pencil and the last page of this project on which to draw their map. Give as much time as the subject needs.

*Part A*. Ask the person what they found most difficult or confusing about the map. What did they think gave them the most difficulty? Did they use any particular feature as a central reference point?

_____

_____

_____

_____

_____

_____

_____

_____

*Part B*. Compare their map with the original and describe the results. For example, did they forget specific buildings, if so which ones? Were all segments of the walkways drawn? Were the distances between buildings distorted, if so how? Were distances near the center of the map distorted differently than those away from the center? Were the shapes of the buildings correct?

_____

_____

_____

_____

_____

_____

_____

_____

_____

_____

_____

Name ______________________________ Course __________ Seat Number __________ Date ________

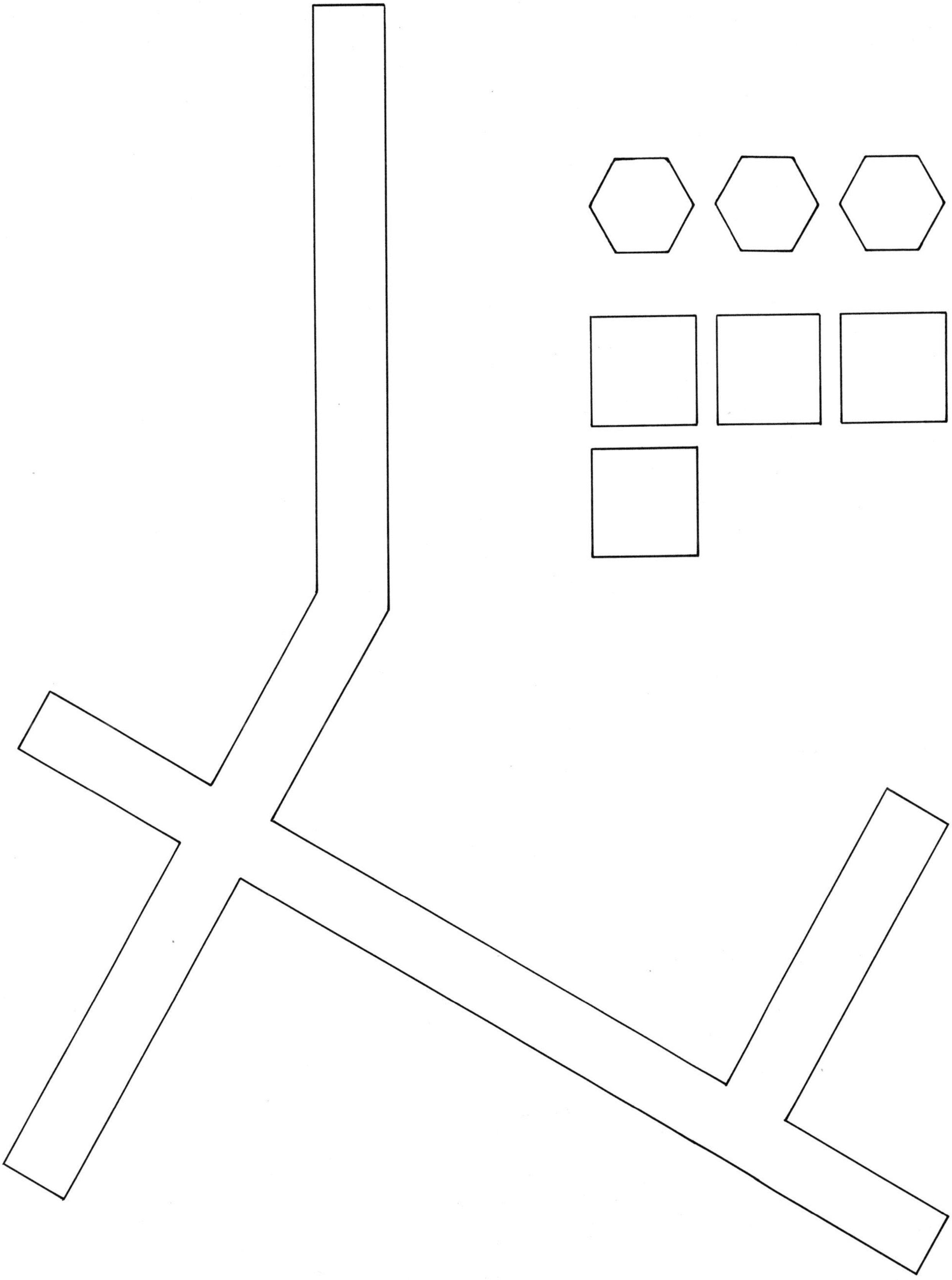

**Figure 1.** Shapes to cut out for construction of the tactual map.

Name ______________________ Course __________ Seat Number __________ Date ________

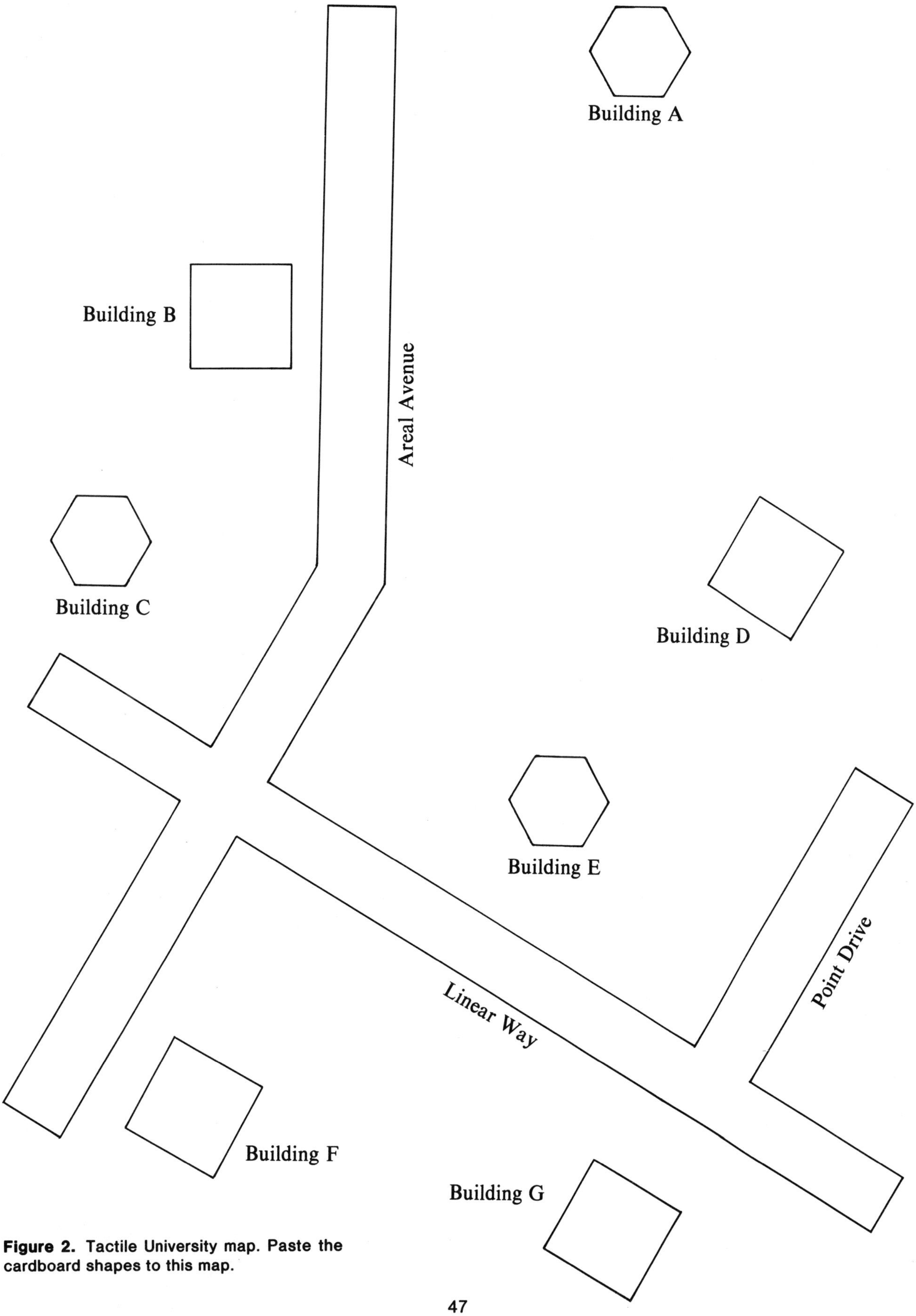
Building A
Building B
Areal Avenue
Building C
Building D
Building E
Point Drive
Linear Way
Building F
Building G

**Figure 2.** Tactile University map. Paste the cardboard shapes to this map.

Paper for drawing cognitive map. Turn in with project.

# Chapter 7

# SPEECH COMMUNICATIONS

**Terms/Concepts:**

Phoneme
Waveform
Spectrum
Sound spectrogram
Intelligibility
Low-pass filter
High-pass filter
Peak clipping
Center clipping
Articulation index

Preferred-octave speech interference level
(PSIL)
Preferred noise criteria curves
Reverberation
Synthesized speech
Analysis-synthesis
Synthesis by rule
Coarticulation
Prosody

**Review Questions:**

1. What are the major components of speech communication systems?
2. Discuss various characteristics of a message that can influence its intelligibility.
3. Discuss the effects on intelligibility of filtering and amplitude distortion.
4. Compare AI, PSIL, and PNC as indices for assessing the effects of noise on speech intelligibility.
5. Summarize the effects of synthesized speech on intelligibility, memory, and preference.

**Activities:**

1. Find a radio, recorder, or record player with bass and treble controls. Play something that has a lot of talking in it, (e.g., a news program). First adjust the bass all the way up and the treble all the way down (low-pass filter). Listen to the quality of speech and the relative intelligibility of the vowel and consonant sounds. Now turn the bass down and the treble up (high-pass filter). Does the speech seem more intelligible? Why? What about the vowel and consonant sounds?
2. Listen to synthesized speech and judge its quality. Some places to hear synthesized speech are telephone company directory assistance services, automobiles, home computer speech synthesizers, and cameras.

Project 7

# THE USE OF THE PREFERRED-OCTAVE SPEECH INTERFERENCE LEVEL (PSIL)

## Reading Assignment

Text: Chapter 7, pages 183–185.

## Purpose

The purpose of this project is to demonstrate the procedure for applying the PSIL to noise situations with an emphasis on the types of inferences that are possible.

## Problem

*Part A.* Compute the preferred-octave speech interference level (PSIL) for the three hypothetical situations given below.

| Midpoint of Octave Band (HZ) | Intensity (db) in Hypothetical Situations | | |
|---|---|---|---|
| | A | B | C |
| 500 | 69 | 37 | 23 |
| 1,000 | 81 | 86 | 30 |
| 2,000 | 75 | 42 | 37 |
| Total | _____ | _____ | _____ |
| Total ÷ 3 = PSIL = | _____ | _____ | _____ |

*Part B.* Are any of the "assumptions" regarding the accuracy of the PSIL violated in either situation listed above? If so, what?

_____

_____

_____

*Part C.* Using Figure 7-9, on page 184 of the text, answer the following questions.

1. In situation A, above, what approximately is the maximum distance people can be apart and have a reliable conversation if they are shouting?

_____ feet apart

2. If two people are working eight (8) feet apart in situation A, would it be possible, difficult, or impossible to communicate?

_____

3. In situation B, if two people are ten feet apart, what level of voice will they have to use in order to have a reliable conversation?

_____ level of voice

*Part D.* Using Figure 7-10 on page 185 of the text, answer the following questions.

1. If the three situations were private offices, what approximately would be the subjective rating of the noise by people in each situation?

Situation A _____

Situation B _____

Situation C _____

2. In which situation(s) A and/or B, would you recommend a telephone operator be located?

Situation _____

*Part E.* Using Table 7-3, page 186 of the text, answer the following questions.

1. What would be two possible uses for Situation B?

Use 1: _____

Use 2: _____

2. Would it be permissible to use Situation C as a:

Movie theater?                    _____

Theater for drama?                _____

Concert hall?                     _____

*Part F.* What do you see as the advantages and disadvantages of using PSIL as a measure of the speech interference of noise?

_____

_____

_____

_____

_____

_____

# Chapter 8

# HUMAN PHYSICAL ACTIVITIES

**Terms/Concepts:**

Stress

Strain

Synovial joints

Cartilaginous joints

Striated muscles

Motor (efferent) nerves

Sensory (afferent) nerves

Motor end plates

Metabolism

Anaerobic

Aerobic

Oxygen debt

Basal metabolic rate

Kilocalorie

Stroke volume

Heart rate recovery curve

Electromyographic (EMG) recordings

Flicker fusion frequency (FFF)

Critical fusion frequency (CFF)

Evoked cortical potential (ECP)

Workload

Work efficiency

Biomechanics

Types of movements of body members

Force platform

Isotonic or isokinetic strength

Isometric strength

Endurance

Simple reaction time

Choice reaction time

Expectancy

Movement time

Static muscular control

Tremor

Maximum permissible limit (MPL)

Action limit (AL)

**Review Questions:**

1. Discuss the metabolic process involved in muscular work.

2. List some measures of gross body activity, local muscular activity, and physiological strain.

3. Discuss some of the things that effect workload in specific work activities.

4. What are two types of reaction time tasks and what are some variables that affect them?

5. Discuss several variables that affect manual materials handling tasks.

**Activities:**

1. Time how long you can hold your textbook in one hand, palm down, with your arm fully extended to the side at shoulder level. Which gives out first, your grip or your shoulder? Try it with your other arm except bend your elbow so that the book is closer to your shoulder. Which side could you hold longer? Why?

2. Perform the movements shown in Figure 8-13, page 214 of the text. Compare the relative range of movements as shown in the figure.

3. Get a piece of stiff cardboard and a marking pen. Stand the cardboard up with the edge facing you, approximately two feet away. Hold the pen in your finger tips with your palm up. Try to draw three straight horizontal lines on the cardboard (it requires an in-out motion). After a short rest, try to draw three straight vertical lines with an up-down motion on the cardboard. Which set of lines appear straighter and less wiggly? How does that compare to the results shown in Figure 8-20, page 224 of the text?

Project 8-A

# WORK PACE AND ENERGY COSTS OF LIFTING

## Reading Assignment

Text: Chapter 8, pages 211–212, 225–229.

## Purpose

The purpose of this project is to illustrate the effect of load weight and lift ranges on energy expenditure in a lifting task.

## Problem

The XYZ company manufactures, among other things, metal ingots weighing 30 lbs. each. As the ingots come off a conveyor 8 inches off the floor, a "loader" man places them on a pallet (8 inches off the floor), one on top of another. He lifts the ingots an average of 12 inches above the coveyor.

The pallet is then moved by forklift to the storage area and placed on a platform 40 inches above the floor. The "storage" man, standing on the floor, then puts the ingots on various shelves 40 to 60 inches above the floor. He also lifts the ingots an average of 12 inches.

*Part A.* The loader lifts one ingot (30 lbs) at a time, an average of 12 inches, (lift range 8–20 in) 1,500 times per hour.

1. Using Figure 8-23 (page 228 of the text) and the formula given below, determine the total energy expenditure (in kcal/h) for this rate of work.

   *Values:*

   E = energy expenditure (Kcal/h) = to be determined
   N = number of lifts/h = 1500
   H = height of lift (ft.) = 1.0
   W = weight of object lifted (lb.) = 30
   C = energy (gram calories/ft. lb.) from Figure 8-23, page 228 of the text = _____

   *Formula:*                           *Computations:*

   $$E = \frac{N \times H \times W \times C}{1000}$$

   Total energy expenditure = _____ kcal/h.

   ÷ 60 = _____ kcal/min.

2. Using Table 8-2 on page 211 of the text, what grade of work does this represent?

   Grade of work = _____

3. Using Figure 8-12 (page 212 of the text) and adopting an energy standard of 4 kcal/min., how much rest per hour would be required for this rate of work?

   Rest required _____ minutes/hour

*Part B.* The company would like to set paces (number of lifts per hour) for the two men so that they were expending a reasonable amount of energy at the task. The text (page 211) suggests, as reasonable, an overall expenditure level of around 4 kcal/min. (240 kcal/h).

1. Using this level and the formula given below, determine the maximum number of lifts per hour that the loader (lift range 0–20 in.) could do and not exceed 240 kcal/h. energy expenditure if he lifted one ingot at a time.

   *Values:*

   E = energy expenditure (kcal/h) = 240
   N = number of lifts/h = to be determined
   H = height of lift (ft.) = 1.0
   W = weight of object lifted (lb.) = 30
   C = energy consumption (gram calories/ft. lb.) from Figure 8-23, page 228 of the text = _____

   *Formula:*                          *Computations:*

   $$N = \frac{1000 \times E}{H \times W \times C}$$

   Maximum number of lifts per hour = _____

2. Determine the maximum number of lifts per hour the storage man (lift range 40–60 in.) could do and maintain a total energy expenditure of 240 kcal/h. if he lifted one ingot at a time.

   *Values:*

   E = energy expenditure (kcal/h) = 240
   N = number of lifts/h = to be determined
   H = height of lift (ft.) = 1.0
   W = weight of object lifted (lb.) = 30
   C = energy consumption (gram calories/ft. lb.) from Figure 8-23, page 228 of the text = _____

   *Formula:*                          Computations:

   $$N = \frac{1000 \times E}{H \times W \times C}$$

   Maximum number of lifts per hour = _____

*Part C.* A foreman notices that the men are picking up only one ingot at a time. He feels they could, without increasing their total energy expenditure, load and store more ingots if they picked up two at a time (60 lbs.). To do this, of course, the men would have to change their rate of lifting.

1. Recalculate for the LOADER (lift range 0–20 in.) the maximum number of lifts he could do and still maintain an overall total energy expenditure of 240 kcal/h. lifting 60 lbs. instead of 30 lbs.

   *Values:*

   E = energy expenditure (kcal/h) = 240
   N = number of lifts/h = to be determined
   H = height of lift (ft.) = 1.0
   W = weight of object lifted (lb.) = 30
   C = energy consumption (gram calories/ft. lb.) from Figure 8-23, page 228 of the text = _____

   *Formula:*                                  *Computations:*

   $$N = \frac{1000 \times E}{H \times W \times C}$$

   Maximum number of lifts/h. = _____

   At two ingots per lift this equals _____ ingots/h.

2. Recalculate for the STORAGE MAN (lift range 40–60 in.) the maximum number of lifts per hour he could do and still maintain an overall total energy level of 240 kcal/h. lifting 60 lbs. instead of 30 lbs.

   *Values:*

   E = energy expenditure (kcal/h) = 240
   N = number of lifts/h = to be determined
   H = height of lift (ft.) = 1.0
   W = weight of object lifted (lb.) = 60
   C = energy consumption (gram calories/ft. lb.) from Figure 8-23, page 228 of the text = _____

   *Formula:*                                  *Computations:*

   $$N = \frac{1000 \times E}{H \times W \times C}$$

   Maximum number of lifts/h. = _____

   At two ingots per lift this equals _____ ingots/h.

*Part D.* Determine the relative efficiency of lifting one versus two ingots at a time for loader and storage man.

1. For the loader, which would be more efficient, i.e., higher number of ingots per hour at the same energy level, if he lifted one ingot at a time or two?

More efficient to lift _____ ingots/lift

2. For the storage man, which would be more efficient, if he lifted one ingot at a time or two?

More efficient to lift _____ ingots/lift

3. Why is the more efficient load (one versus two ingots) not the same for the loader and storage man?

_____

_____

_____

Project 8-B

# WORK POSTURE AND ENERGY EXPENDITURE

## Reading Assignment

Text: Chapter 8, pages 206–208.

## Purpose

The purpose of this project is to illustrate the effect of posture on energy expenditure for a relatively simple task.

## Problem

In this project you are asked to replicate part of the experiment performed by Vos (1973) and discussed in the text on page 206. The task consists of picking up 160 small objects off the ground using two different work postures. Pulse rate will serve as the measure of relative energy expenditure.

Procedure:

1. For this project, you will need 160 paper clips or small pieces of paper, a small bowl or cup, and a watch with a second hand.
2. Either you or a friend can serve as the subject. Anyone with a heart or respiratory condition should sit this one out.
3. Place the 160 objects on a floor in the pattern shown in Figure 1. You will need an area 10 $\times$ 4 feet in size.
4. Have the subject rest in a chair for 10 minutes.
5. After the 10 minutes is up determine the subject's pulse rate. An easy way to count the pulse is in the wrist. Have the subject turn their right hand, palm up and press your fingers over the radial artery on the thumb side of the wrist (see Figure 11-1, page 304 of the text). Count the pulse for exactly 15 seconds and multiply it by 4 to get beats per minute. Record this in Table 1 in the "Squatting-Rest" box.
6. Hand the subject the bowl and have them squat as shown in Figure 8-7 on page 208 of the text. The subject is to pick up all the objects in rows 1, 2, 3, and 4 as shown in Figure 1 and then come back picking up the objects in rows 5, 6, 7, and 8. As the objects are picked up, they are placed in the bowl. The subject must maintain the basic posture as they move to pick up the objects.
7. When the subject completes the task, take their pulse again for 15 seconds. Multiply by 4 and record in Table 1.
8. Have the subject sit in a chair and rest. Every 60 seconds for 3 minutes take their pulse (15 s, multiply by 4), and record each of the three pulse rates in Table 1.
9. Have the subject rest for at least another 10 minutes. During this time, place the 160 objects on the floor in the same pattern as before and shown in Figure 1. At the end of the rest period take the subject's pulse (15 s, multiply by 4) and record it in Table 1 "Bending: no arm support—Resting".
10. Hand the subject the bowl and have the subject bend over at the waist without supporting their arm as shown in Figure 8-7. The subject is to pick up the objects as he or she did before, i.e., rows 1, 2, 3, and 4 and then coming back, 5, 6, 7, and 8.
11. Take the subject's pulse at the end of the task as before and record in Table 1.
12. Have the subject sit in a chair and rest. Take their pulse every 60 seconds for 3 minutes (15 s, multiply by 4). Record the three pulse rates in Table 1.
13. Plot the pulse rates for the two postures in Figure 2.

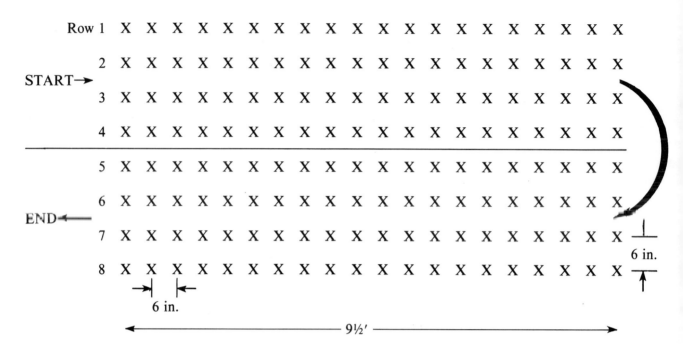

**Figure 1.** Pattern for laying out objects on the floor and picking them up.

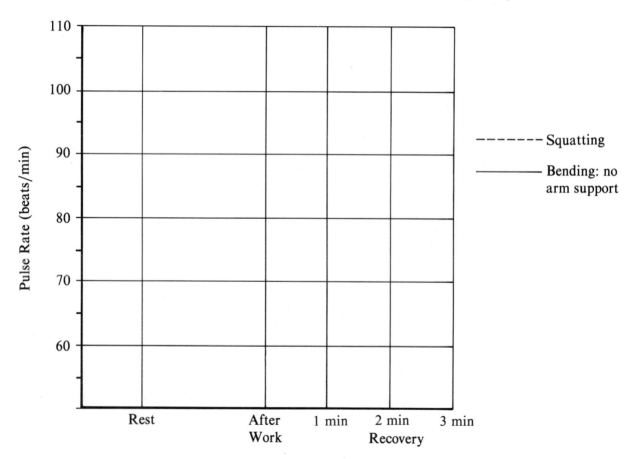

**Figure 2.** Graph of pulse rate data.

**TABLE 1**
Pulse Rates (beats per minute)

| | Rest | After Work | 1 min | Recovery 2 min | 3 min |
|---|---|---|---|---|---|
| **Squatting** | | | | | |
| **Bending: no support** | | | | | |

*Part A.* How did your results compare to those reported in Figure 8-7, page 208 of the text?

_____

_____

_____

_____

*Part B.* Why do you think it is more efficient to squat then to bend at the waist?

_____

_____

_____

_____

*Part C.* Brouda used heart rate recovery curves (page 203 of the text) as a measure of energy consumption. How did the recovery data you collected reflect different energy expenditures after using the different postures?

_____

_____

_____

_____

_____

_____

Project 8-C

## APPLICATION OF NIOSH LIFTING LIMIT FORMULA

### Reading Assignment

Text: Chapter 8, pages 225–230; Appendix C, page 646.

### Purpose

The purpose of this project is to illustrate the use of the NIOSH recommended lifting limit formula to determine a safe lifting weight and to demonstrate what effect redesigning a work site can have on the recommended limits.

### Problem 1

As part of the manufacturing process at the ABC Co., 25 lbs. sacks of chemicals have to be dumped into a storage bin. The sacks are delivered to the work site by a floor-level conveyor. The sacks, therefore are 10 in. off the floor. The worker stands at the end of the conveyor and lifts the sacks by hand and dumps them into the storage bin 45 in. above the floor. The conveyor is positioned behind pipes running along the floor. This makes it impossible for the worker to get closer than 24 in. to the sacks when lifting them. The task of filling the bin takes less than one hour and the worker lifts an average of 4 lifts per min.

The company discovers that if they purchase the chemicals in 50 lbs. sacks they can save a considerable amount of money. They are concerned, however, that the heavier sacks may pose a danger to the worker filling the bin.

*Part A:* Using the formula in Appendix C, calculate the Action Limit (AL) in pounds (lbs.) for this task.

*Values:*

H = horizontal distance from ankles to object (in.)   = _____
V = vertical height of object at start of lift (in.)   = _____
D = vertical travel distance of lift (in.)   = _____
F = average frequency of lift (lifts/min)   = _____
F max = maximum frequency from chart in Appendix C   = _____

*Factors:*

Horizontal location   = 6/H   = _____
Vertical location   = $1-(0.01 \times |V-30|)$   = _____
Distance traveled   = 0.7 + 3/D   = _____
Frequency of lift   = 1−F/F max   = _____

*Formula:*

AL = 90 × (horizontal location factor) × (vertical location factor) × (distance traveled factor) × (frequency of lift factor)

AL = _____ lbs.

*Part B:* Using the formula in Appendix C, calculate the Maximum Permissible Limit (MPL) for this task and determine if it is safe to lift the larger 50 lbs. sacks.

*Formula:*    MPL = 3 × AL

MPL = _____ lbs.

Is it safe to lift the 50 lbs. sacks? _____

## Problem 2

The company human factors specialist is called in to analyze the task and work site described in Problem 1. The specialist recommends that the conveyor be raised so that the sacks are 35 in. off the floor. In addition, by raising the conveyor off the floor, the conveyor can be extended over the pipes thus allowing the worker to get as close as 12 in. to the sacks when lifting.

*Part A:* Recalculate the Action Limit given the improvement recommended by the human factors specialist. (Assume that the worker would work at the same pace of 4 lifts/min.)

*Values:*

H = horizontal distance from ankles to object (in.)      = _____
V = vertical height of object at start of lift (in.)      = _____
D = vertical travel distance of lift (in.)      = _____
F = average frequency of lift (lifts/min)      = _____
F max = maximum frequency from chart in Appendix C  = _____

*Factors:*

Horizontal location   = 6/H      = _____
Vertical location    = 1−(0.01 × |V−30|) = _____
Distance traveled    = 0.7 + 3/D    = _____
Frequency of lift    = 1−F/F max    = _____
*Formula:*

AL = 90 × (horizontal location factor) × (vertical location factor) × (distance traveled factor) × (frequency of lift factor)

AL = _____ lbs.

*Part B:* Calculate the Maximum Permissible Limit (MPL) given the recommendation of the human factors specialist and determine if the improvement would make it safe to lift the larger 50 lbs. sacks.

*Formula:*    MPL = 3 × AL

MPL = _____ lbs.

Is it now safe to lift the 50 lbs. sacks? _____

*Part C:* Which aspect of the recommendation, increasing the vertical height of the sacks or allowing the worker to move closer to the sacks had the greatest effect on the AL and MPL?

Check one:

Increased vertical height of load      _____
Decreased horizontal distance to load   _____

# Chapter 9

# HUMAN CONTROL OF SYSTEMS

**Terms/Concepts:**

Compatibility
Spatial compatibility
Movement compatibility
Population stereotypes
Warrick's principle
Scale-side principle
Clockwise-for-increase principle
Target
Course
Step, ramp, and sine inputs
Follower or cursor
Controlled element
Pursuit display
Compensatory display
Control order
Zero order control
First order control

Second order control
Bandwidth
Anticipation
Preview
Response lag
Control system lag
Display system lag
Transmission time lag
Exponential lag
Sigmoid time lag
C/R ratio
Specificity of displayed error
Paced versus self-paced tracking
Aiding
Predictor display
Quickening

**Review Questions:**

1. Define and give examples of spatial and movement compatibility.

2. Give an example of a pursuit tracking task and a compensatory tracking task and discuss the advantages and disadvantages of each.

3. Discuss some factors that influence tracking performance.

**Activities:**

1. Draw the four examples shown in Figure 9-4, page 240 of the text, on separate pieces of paper. Ask ten people which direction (clockwise or counterclockwise) they would turn each knob to move the pointer to 15. Compare your results with those given in the text.

2. Go to an arcade where they have electronic games. Look at the shooting and driving games and for each determine if it is a compensatory or pursuit display; whether there is preview of the track; whether it is paced or self-paced; and whether anticipation is possible. Try a few of the games and see if you can detect any time lags in them.

Project 9

## COMPATIBILITY RELATIONSHIPS

### Reading Assignment

Text: Chapter 9, pages 234–243.

### Purpose

The purpose of this project is to illustrate some practical applications of the compatibility relationships discussed in the text. An effort has been made to include situations which differ in the degree to which "population stereotypes" can be employed to solve the problem. Various types of compatibility are illustrated.

### Problem 1

Indicate the recommended direction of movement for the duplicating machine's crank by adding an arrowhead to the arc through the handle to indicate the desired direction of movement.

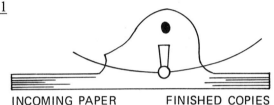

INCOMING PAPER      FINISHED COPIES

### Problem 2

Indicate the recommended direction of movement of the focusing knob to move the lens of the camera in the direction indicated. Adding an arrowhead to the arc below the knob.

FOCUSING KNOB

DESIRED DIRECTION OF LENS MOVEMENT

### Problem 3

*Part A.* Indicate the direction of movement for each sink's faucets that will turn it on. Add an arrowhead to the arc over each faucet.

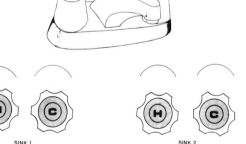

*Part B.* Find two actual sinks, in a kitchen or bathroom, and record the directions of movement required to turn them on.

SINK 1          SINK 2

### Problem 4

In which direction would you turn the knob to move the pointer to 15? Add an arrow head to the arc under the knob.

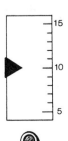

### Problem 5

The following is a keyboard used to enter the numerals 0 to 9 with the fingers of both hands. Indicate how you would assign the digits 0 to 9 to the buttons.

### Problem 6

You're flying in an aircraft cockpit, you hear the statement "LEFT WING DOWN." How should the pilot act to change his roll? Check one.

Lower left wing _____

Raise left wing _____

### Problem 7

In underground coal mining, the miner often sits on the side of the machine and operates the controls. In what direction, up or down, would you move the control to cause the machine to move in the direction of the arrow?

Circle one: UP   DOWN

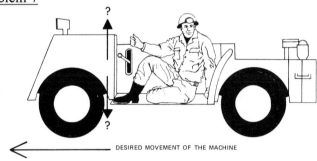

DESIRED MOVEMENT OF THE MACHINE

### Problem 8

Which way would you turn the key in the deadbolt lock to unlock the door. Add an arrowhead to the arc over the lock.

### Problem 9

The knob, with numbers and directional arrows, shown to the right was actually used on an office copier to control the darkness of the copies to be made. In which direction, A or B, would you turn the knob to make the copies darker?

Circle one: A   B

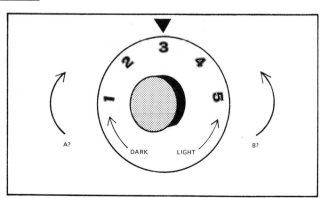

70

## Problem 10

The following are concentrically mounted knobs such as shown in Figure 10-13, page 277 of the text. Indicate which knob controls each display window by placing the appropriate letter in the window.

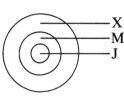

## Problem 11

On the four-lane divided highway pictured below, indicate which is the outside lane.

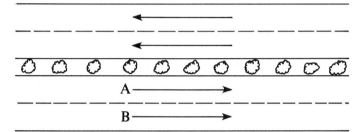

Check one:

Lane A _____

Lane B _____

## Problem 12

In which direction, A or B, would you move the lever to move the pointer to the right?

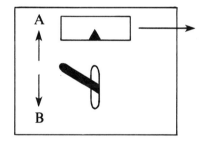

Check one:

A (up) _____

B (down) _____

## Problem 13

In which direction, A or B, would you move the lever to move the pointer to the left?

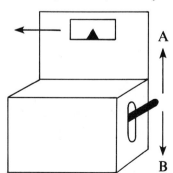

Check one:

A (up) _____

B (down) _____

## Problem 14

The hand-held calculator shown to the right needs the arithmetic functions add ($+$), subtract ($-$), multiply ($\times$), divide ($\div$), and equals ($=$) assigned to the five blank keys. Which key should control each function? Write in the function symbols on the appropriate keys.

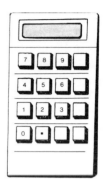

## Problem 15

The panel to the right is an elevator light panel used in a ten-story building. As the elevator moves from floor to floor, the appropriate light comes on. How would you assign the floor numbers to the lights on the panel. Write the floor numbers on the appropriate lights.

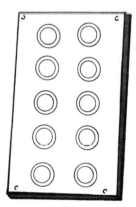

## Problem 16

In which direction would you turn the knob to move the pointer to 15? Add an arrow head to the arc under the knob.

## Problem 17

The computer screen to the right shows one portion of an ordered list of numbers (1–200). Which button would you use to view the number "3" on the screen?

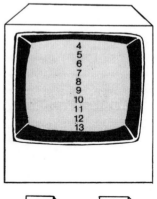

A

B

Circle one: A    B

# Chapter 10

# CONTROLS

**Terms/Concepts:**

Discrete information
Continuous information
Cursor positioning
Detent positioning
Types of control-coding methods
Control-response ratio
Gain or sensitivity
Free position or isotonic controls
Pure force or isometric controls
Elastic resistance
Static friction
Coulomb friction
Viscous damping
Inertia
Difference limen
Deadspace
Backlash

Displacement joystick
Spring-return joystick
Fitts' law
Chord keyboard
Sequential keyboard
QWERTY keyboard
Dvorak or simplified keyboard
Hysteresis
Membrane keyboard
Touch screen
Touch pad
Light pen
Digitizing tablet
Mouse
Teleoperator
Types of speech recognition systems

**Review Questions:**

1. Discuss the advantages and disadvantages of the various control coding methods presented in the text.

2. What is C/R ratio, what is its effect on gross movement and adjustment time, and how is the optimum C/R ratio determined?

3. Describe and explain the advantages and disadvantages of the various types of resistance discussed in the text.

4. Describe and explain the advantages and disadvantages of the various types of cursor positioning devices discussed in the text.

5. What are the major types of speech recognition systems and what are some of the human factors issues involved in their use?

**Activities:**

1. Examine the stalk controls on several late model cars. The stalk control is mounted on the steering column and often contains multiple controls such as shown in Figure 10-6, page 266 of the text. For each model of car make a list of the functions controlled on the stalk and indicate how each is activated. Compare the various designs. What are the advantages and disadvantages of each? Which would you recommend and why?

2. Go to a video arcade and play various games that involve different types of controls such as joysticks (isometric or isotonic), trackballs, and cursor keys. Which type of controls did you prefer for the various types of tasks involved in the games? Why?

3. Go to a computer store and experiment with various types of cursor positioning controls. Try simple cursor positioning and pointing. Which type of controls did you prefer? Why? Try to draw something on the screen using the various controls. Which did you prefer for this type of task? Why?

Project 10

## MOVEMENT TIME FOR OPERATING FOOT PEDALS

### Reading Assignment

Text: Chapter 10, pages 280–283.

### Purpose

The purpose of this project is to illustrate the effect of movement distance, pedal width, and shoe width on reciprocal and single-movement times in the operation of foot pedals.

### Problem

An engineer for Brimark Innovations was designing a device that required rapid foot movements between two pedals for its operation. (We will suspend judgment on the desirability of the design concept itself.) The engineer can place the pedals either 18 in. or 9 in. apart (center-to-center) and can use either 2 in. or 4 in. wide pedals. Unfortunately, using the larger pedals or placing them 9 in. apart will entail additional costs for tooling, parts, and assembly time. The engineer comes to you to find out about the effects on movement time (both reciprocal and single-movement) of using the various inter-pedal distances and pedal sizes.

In addition, the device will be operated by men with an average shoe width of 4 in. and by women with an average shoe width of 3 in. The engineer is also interested in the effect, if any, the different shoe widths will have on movement times.

*Part A:* Using the formulas on pages 282 and 283 of the text, compute the index of difficulty (ID), reciprocal movement time (RMT), and single-movement time (SMT) for the combinations of pedal distance (A), pedal width (W), and shoe width (S) discussed above. Use Table 1 to record your results. Use the following conversion table to convert $A/(W + S)$ to $\log_2 [A/(W + S)]$. Interpolate between values if necessary.

| A/(W+S) | Log₂ A/(W+S) | A/(W+S) | Log₂ A/(W+ |
|---------|--------------|---------|------------|
| 1.0 | 0.000 | 2.6 | 1.378 |
| 1.1 | 0.137 | 2.7 | 1.433 |
| 1.2 | 0.263 | 2.8 | 1.485 |
| 1.3 | 0.378 | 2.9 | 1.536 |
| 1.4 | 0.485 | 3.0 | 1.585 |
| 1.5 | 0.585 | 3.1 | 1.632 |
| 1.6 | 0.678 | 3.2 | 1.678 |
| 1.7 | 0.765 | 3.3 | 1.722 |
| 1.8 | 0.848 | 3.4 | 1.765 |
| 1.9 | 0.926 | 3.5 | 1.807 |
| 2.0 | 1.000 | 3.6 | 1.848 |
| 2.1 | 1.070 | 3.7 | 1.887 |
| 2.2 | 1.137 | 3.8 | 1.926 |
| 2.3 | 1.202 | 3.9 | 1.963 |
| 2.4 | 1.263 | 4.0 | 2.000 |
| 2.5 | 1.322 | | |

**Table 1.**

Effect of inter-pedal distance, pedal width, and shoe sole width on movement times.

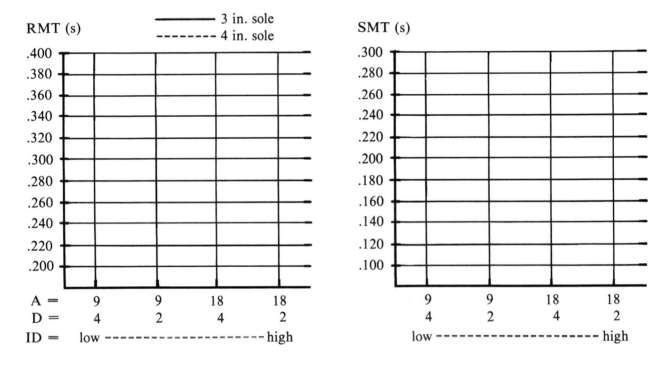

**Part B:** Graph the RMTs and SMTs in Figure 1. Use a solid line for the 3 in. sole data and a dashed line for the 4 in. sole width data.

*Part C:* In general, which variable has the greatest effect on movement times, movement amplitude (A) or pedal width (W)?

_____

*Part D:* Under what conditions did shoe width have the greatest effect?

_____

_____

The least effect? _____

_____

*Part E:* The engineer asks whether the difference in RMT and SMT between the best and worst conditions justify the additional costs for the better design. What additional information would you need before you could answer the question?

_____

_____

_____

_____

_____

# Chapter 11

# HAND TOOLS AND DEVICES

**Terms/Concepts:**

Carpal tunnel
Pisiform bone
Ulna bone
Radius bone
Palmar flexion
Dorsiflexion
Ulnar deviation
Radial deviation
Humerus

Bicep
Tenosynovitis
Carpal tunnel syndrome
Epicondylitis or tennis elbow
Ischemia
Trigger finger
Grip axis
Vibration-induced white finger

**Review Questions:**

1. Discuss the consequences of working with a bent wrist.
2. What are some of the human factors principles of hand tool design?
3. List various techniques and methods used to evaluate hand tools and related devices.

**Activities:**

1. Go to a hardware store and compare the design of various models of a particular type of hand tool, such as hand drills, circular saws, pliers, or electric sanders. Compare such things as handle diameter, grip axis, shape, position of on-off switch, placement of guards, accommodation of left handers, and other relevant human factors features.
2. Get one or more pairs of gloves. If possible, get them in different styles or materials. Determine how fast you can tie your shoes with the gloves on, and with no gloves on. Try picking up small objects such as paper clips, coins, etc. What features of the gloves seem to hinder performance?
3. Try various types of toothbrushes, including the Reach toothbrush. What features do you like and dislike?
4. Go to a camera store and compare various disc cameras including the Kodak models. Which do you like best? Why?

Project 11-A

## REDESIGN OF A SOLDERING IRON

### Reading Assignment

Text: Chapter 11, pages 303–313.

### Purpose

The purpose of this project is to afford the student an opportunity to apply information gained from the text to the practical problem of tool design.

### Problem

In a small manufacturing plant, a particular soldering iron is used to solder connections on a large vertical panel. Several problems have been uncovered by observing and interviewing the operators, and reading accident reports.

First, when holding the iron as in Figure 1, it is difficult to contact the desired electrical connection perpendicularly. The tendency is to use the side of the iron's point, rather than the tip of the point. This results in poor quality work.

Second, many operators hold the iron too tightly causing the knob on the butt of the handle to be driven in the palm of the hand, as shown in Figure 1. This in turn causes irritation and pain in the hand. It is felt that an operator grips the hot iron too tightly due to the fear of the hand sliding forward onto the hot iron.

Third, many accidents are caused by the electrical cord becoming entangled so the iron is jerked back into the operator's hand as it is moved forward to the work position.

Fourth, there have been complaints of wrist pain among the workers.

*Part A.* In the space below Figure 1, sketch a redesigned soldering iron which you feel will eliminate or reduce the problems outlined above. Point out any special features which have been incorporated.

*Part B.* Justify your design. How will it eliminate or reduce the problems discussed?

_____

_____

_____

_____

_____

_____

_____

_____

_____

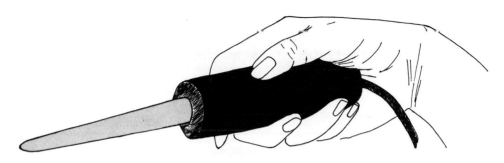

**Figure 1.** Common method of using the soldering iron.

Your design: (Point out any special features which you incorporate.)

Project 11-B

## EVALUATING A NEW BASEBALL BAT

### Reading Assignment

Text: Chapter 11, pages 305–309; 320–326.

### Purpose

The purpose of this project is to permit the student to design the evaluation of a hand tool using laboratory and field techniques.

### Problem

Your text discusses the use of bent handles for tools and sports equipment as patented by John Bennett (pages 306–307 of the text). As indicated, Mr. Bennett has developed a softball bat based on the bent handle concept. Your task in this project is to design a comprehensive evaluation of the new softball bat. The bat is proported to provide a more powerful and accurate swing and to allow the hitter better control of the ball. The evaluation should include laboratory tests, field tests, and user preference data.

*Part A.* Describe laboratory tests for the new bat. Specifically, list the dependent measures you would use, how you would collect the data, and what the subjects would be required to do. Provide as much detail as you can. (Use extra paper if necessary.)

_____

_____

_____

_____

_____

_____

_____

_____

_____

_____

_____

_____

*Part B.* Describe the field tests you would conduct. What measures would you take and how would you conduct the study. Provide detail and use extra paper if necessary.

_____

_____

_____

_____

_____

_____

_____

_____

_____

_____

*Part C.* What questions would you ask users? In what format would you require them to respond?

_____

_____

_____

_____

_____

_____

_____

_____

_____

# Chapter 12

# APPLIED ANTHROPOMETRY AND WORK SPACE

**Terms/Concepts:**

Engineering anthropometry
Structural (static) anthropometry
Functional (dynamic) anthropometry
Somatography
Principles for applying anthropometric data
Work-space envelopes
Kinetosphere

Strophospheres
Normal and maximum arm reach areas
Lumbar section of the spine
Kyphosis
Lordosis
Ischial tuberosities (sitting bones)

**Review Questions:**

1. Discuss the three principles of applying anthropometric data.

2. What are some factors that would influence the recommended work surface height?

3. What are the general principles (not the design values) for seat design related to seat height, seat depth, seat back position, etc.?

4. Discuss the various horizontal work surface areas proposed by Barnes, Farley, and Squires.

**Activities:**

1. Measure the heights of various work surfaces such as kitchen counters, bathroom counters, work benches, grocery store checkout counters, desks, etc. How do they compare with the information presented in the text?

2. Mark off the normal and maximum working areas shown in Figure 12-8, page 344, on a desk. Try working within these limits.

3. Find a computer terminal work station. Compare its dimensions with those shown in Figure 12-21, page 358. What features are not in accordance with the recommendations shown in the text?

4. Observe several people using computer terminals. Compare their posture to the posture shown in Figure 12-20, page 358. How do the postures compare?

Project 12

## ANTHROPOMETRY AND VEHICULAR DESIGN

Reading Assignment

Text: Chapter 12, pages 331–338.

Purpose

The purpose of this project is to illustrate the particular body dimension important in determining various design features of motor vehicles.

Problem

The text, page 338, indicates that the first step in the application of anthropometric data to specific design problems is to determine the body dimension important in the design. Figure 1 is a stylized side view of the interior of a truck cab and a front view of the drivers' seat. The numbered dimensions correspond to those listed below. From Figure 2, select the body dimension which would be critical for determining the cab dimension distances listed below. Record the body dimension number in the space provided below. For each cab dimension, indicate whether the 5th or 95th percentile body dimension should be the design value.

| Cab Dimension (from Figure 1) | Important Body Dimension (number from Figure 2) | Design value 5th or 95th |
|---|---|---|
| 1. Distance from seat to roof. | _____ | _____ |
| 2. Distance from top of foot pedals to lower edge of steering wheel. | _____ | _____ |
| 3. Horizontal distance from lower edge of steering wheel to seat back. | _____ | _____ |
| 4. Vertical distance from lower edge of steering wheel to floor. | _____ | _____ |
| 5. Distance between dashboard and seat back. | _____ | _____ |
| 6. Distance from steering wheel rim to directional signal. | _____ | _____ |
| 7. Width of cab seat. | _____ | _____ |
| 8. Seat depth. | _____ | _____ |
| 9. Width of seat back. | _____ | _____ |
| 10. Height of seat front above floor. | _____ | _____ |

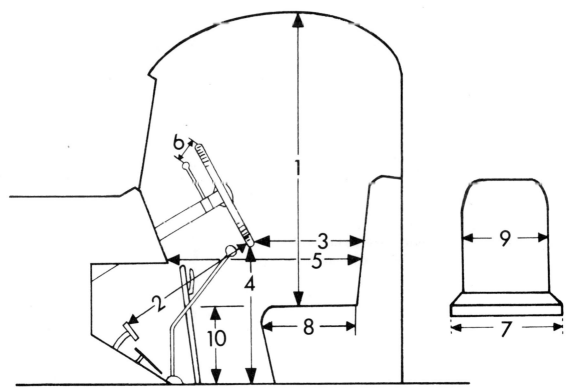

**Figure 1.** Cab dimensions.

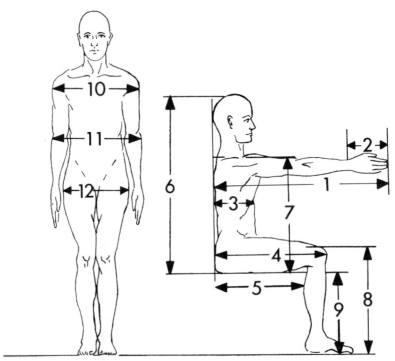

1.  Arm Length.
2.  Hand Length.
3.  Abdomen Depth.
4.  Buttock-knee Length.
5.  Buttock-popliteal Length.
6.  Sitting Height.

7.  Shoulder Height.
8.  Knee Height.
9.  Popliteal Height.
10. Shoulder Breadth.
11. Elbow-to elbow Breadth.
12. Seat Breadth.

**Figure 2.** Body measurements.

# Chapter 13

# PHYSICAL SPACE AND ARRANGEMENT

**Terms/Concepts:**

Principles of arranging components
Control accessibility index
Links
Link table

Adjacency layout diagrams
Spatial operational-sequence diagrams
Linear programming
Isoresponse times

**Review Questions:**

1. Discuss the principles of arranging components.
2. Discuss the types of data used in link analysis and the methods used to display such information.

**Activities:**

1. Observe someone preparing dinner in a kitchen. Count the number of times the person moves from each component to every other component (e.g., stove, sink, can opener, refrigerator, pantry, table) as they prepare the meal. Construct an adjacency layout diagram.

2. Make a diagram of your bedroom and construct a spatial operational-sequence diagram of your movements in the morning when you get up and get dressed.

Project 13

# ARRANGEMENT OF DISPLAYS BY PHYSICAL SIMULATION

## Reading Assignment

Text: Chapter 13, pages 363–375.

## Purpose

The purpose of this project is to illustrate the principles and tradeoffs involved in arranging dials on a console.

## Description

An operator in a petroleum processing plant must watch and report on the condition specified by six dials on his work station console during a complex switching task.

Figure 1 is the configuration of the present console. The link values indicate the frequency with which the eye shifts from one display to another (expressed as percent of eye movements made). Values less than 2% are omitted. Table 1 indicates for each dial, the proportion of time spent on each dial (expressed as percentage of total time), and the importance of each dial as rated by the operator (1 = most important, 6 = not important).

## Problem 1

On pages 367–368 of the text, several ways are discussed for combining index values, such as those given in Table 1, into a composite or combination index of priorities.

*Part A.* For the data in Table 1, combine the "time spent" and "importance" indices into a composite index using addition and multiplication. Convert the "proportion of time spent" on each dial to a rank (1 = most time, 6 = least time), and use the following table to record the data and compute the composite indices.

| Dial | Proportion of Time : Rank (1) | Importance Rank (2) | Composite Index by Addition (1) + (2) | Composite Index by Multiplication (1) × (2) |
|------|-------------------------------|---------------------|---------------------------------------|---------------------------------------------|
| A | _____ | 4 | _____ | _____ |
| B | _____ | 2 | _____ | _____ |
| C | _____ | 3 | _____ | _____ |
| D | _____ | 6 | _____ | _____ |
| E | _____ | 5 | _____ | _____ |
| F | _____ | 1 | _____ | _____ |

*Part B*. What differences, if any, exist between the two methods of combining indices?

_____

_____

_____

*Part C*. Which composite index would you recommend using? Why?

_____

_____

_____

## Problem 2

*Part A*. Based on the information in Figure 1 and Table 1, rearrange the dials in a more optimum configuration. Assume that:

1. Neither the size, orientation, nor face of the dials can be altered.
2. The controls that the operator uses, are separate from the console and will be rearranged *after* the dials are rearranged (i.e., do not be concerned with controls or compatibility).
3. The operation requires more than simple "check reading."
4. The specific sequence in which the operator views the dials is variable and depends on the condition of the other dials (i.e., no specific sequence can be described).
5. The solid line "oval" in Figure 2 represents the preferred area for placing dials, but the **size of the paper represents the total space available to place the dials.**

The last page of this project can be detached and the dial shapes cut out. Arrange these cutouts on Figure 2 and, either trace or attach them to Figure 2 in an optimum configuration. Figure 2 and the dials on the last page represent a one-quarter scale at a viewing distance of approximately 23 in.

*Part B*. Defend your arrangement of the dials. Note any tradeoffs between arrangement principles you made.

_____

_____

_____

_____

_____

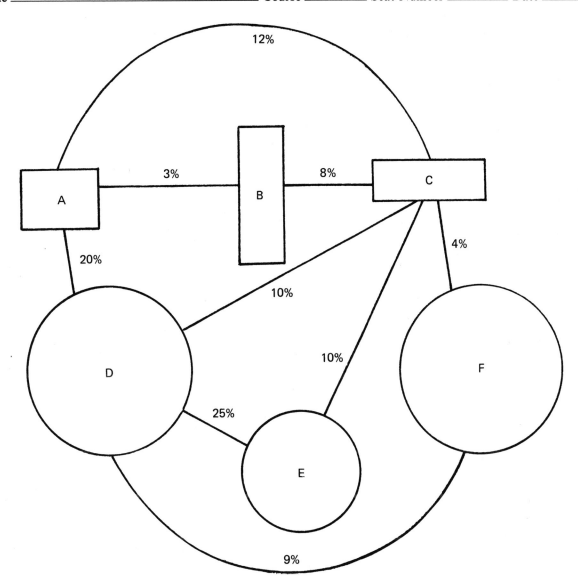

**Figure 1.** Present configuration of the dials with link values indicating the frequency with which the eye shifts from one display to another.

## TABLE 1
Data Required to Rearrange the Dials in Figure 1

| Dial | Proportion of Time Spent | Importance |
|------|--------------------------|------------|
| A | 7% | 4 |
| B | 3 | 2 |
| C | 30 | 3 |
| D | 45 | 6 |
| E | 10 | 5 |
| F | 5 | 1 |
| | 100 | |

Name ______________________ Course __________ Seat Number __________ Date __________

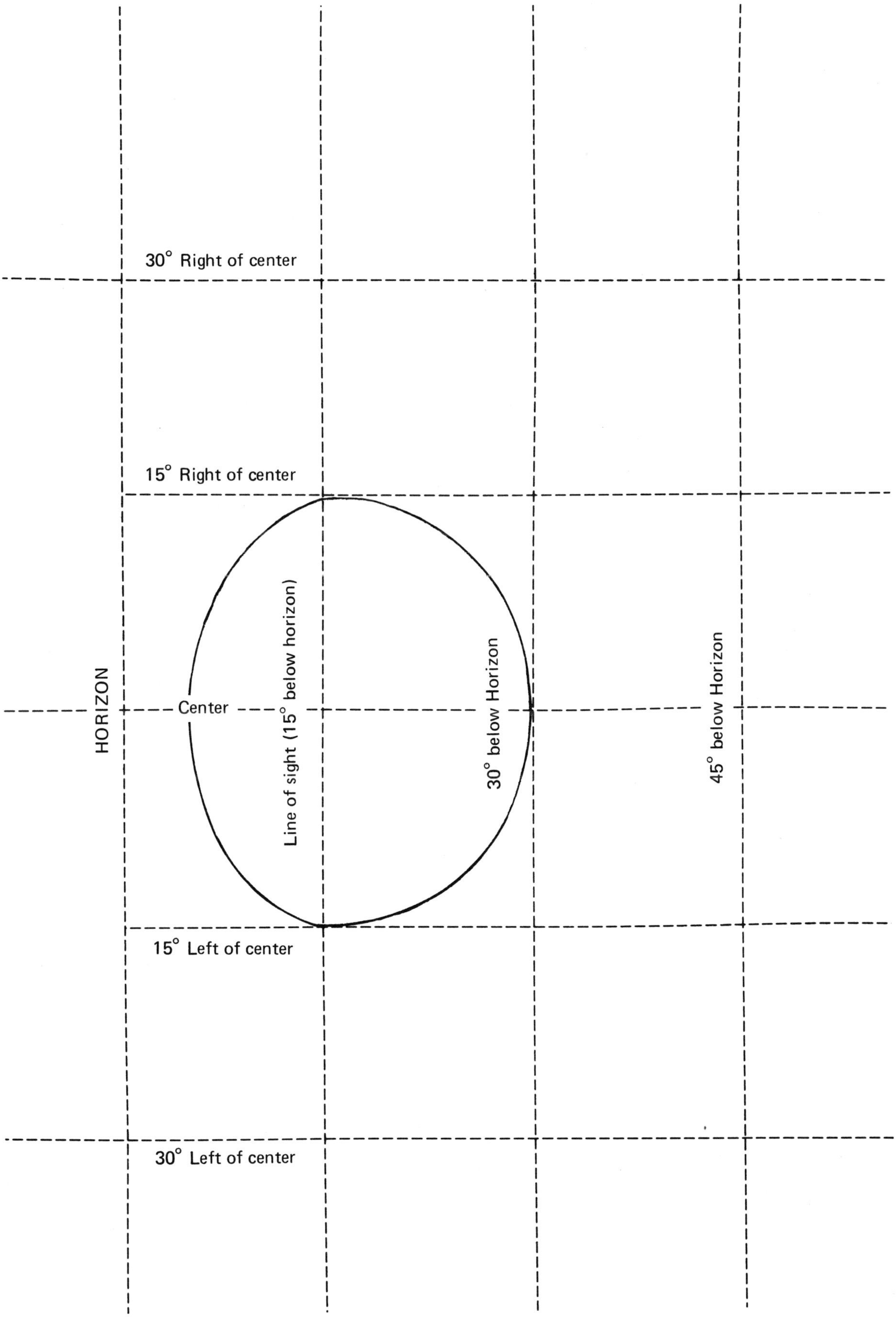
30° Right of center
15° Right of center
HORIZON
Center
Line of sight (15° below horizon)
30° below Horizon
45° below Horizon
15° Left of center
30° Left of center

**Figure 2.** Space on which dials are to be arranged. The solid oval represents the preferred area for placing dials. The size of the paper represents the total space available to place the dials.

Name ______________________ Course ________ Seat Number ________ Date ________

Cut out and use for
rearrangement in
Figure 2.

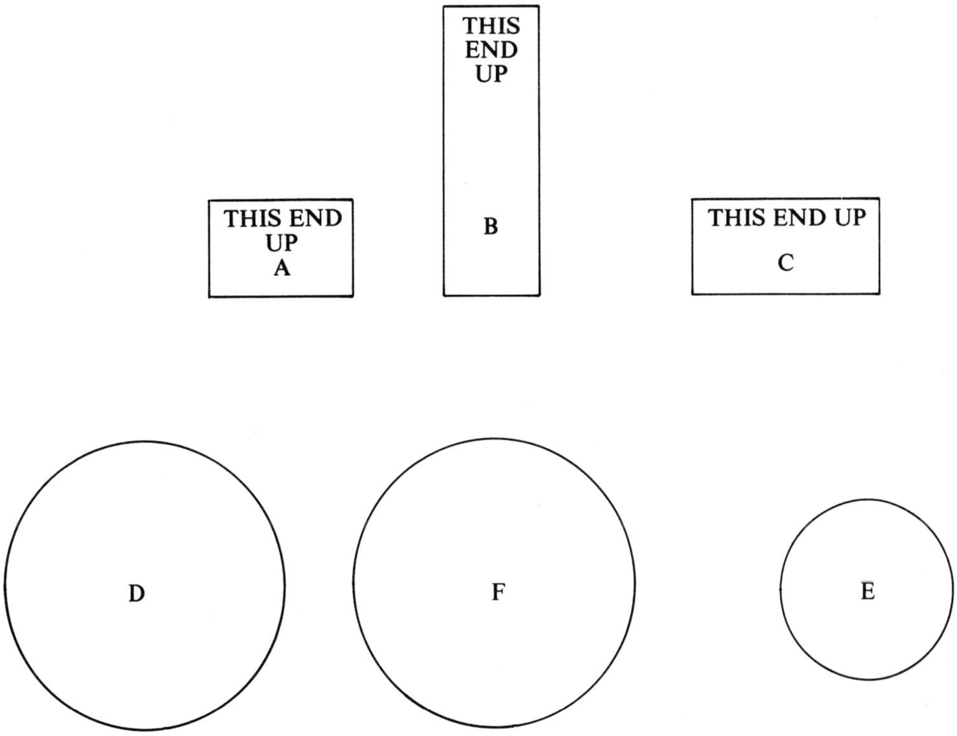
THIS END UP
A
THIS END UP
B
THIS END UP
C
D
F
E

# Chapter 14

# ILLUMINATION

**Terms/Concepts:**

Light
Photopic vision
Scotopic vision
Purkinje effect
Incandescent bodies
Luminescent bodies
White light
Color temperature
Natural color
Hue
Value or luminance
Saturation or chroma
Color cone
CIE colorimetric system
Color coordinates
Chromaticity
Luminous flux
Luminous intensity
Illuminance
Inverse-square law
Illuminance
Reflectance
Lamp
Luminaire

Color rendering index
Lamp efficacy
Visibility
Visibility reference function (VL 1)
Contrast-reducing visibility meter
Equivalent contrast
Effective VL
Transient adaptation
Luminance ratio
Glare
Direct glare
Reflected glare
Types of reflected glare
Discomfort glare
Borderline between comfort and discomfort (BCD)
Discomfort glare rating (DGR)
Visual comfort probability (VCP)
Disability glare
Phototropism
Methods of reducing glare
Methods for reducing screen reflections
Antireflection techniques

**Review Questions:**

1. Discuss the properties of color and their relation to the color cone and CIE chromaticity diagram.

2. Discuss the relationship between luminous intensity, illuminance, and luminance using the concepts of reflectance and the inverse-square law.

3. What are some variables that affect the visibility of a task?

4. Discuss the logic behind Blackwell's VL procedure for measuring task visibility and the criticisms leveled against it.

5. What are some general conclusions regarding the effects of lighting on performance?

6. Why is it not always wise to supply as much illumination as possible in an area?

7. Distinguish between the various types of glare.

8. What are some of the visual changes that accompany old age and what are the implications of these changes for designing lighting environments for the elderly?

9. Discuss illuminance levels, luminance ratios, and screen reflections within the context of video display terminal usage.

**Activities:**

1. Duplicate the experiment shown in Figure 14-15, page 415 of the text. Use a 100 watt light bulb as the glare source and a page of the textbook as the target. Set the book at eye level and try to read the words with the glare source at various heights above the book. How high must you raise it before you can read the page? How high must you raise it before it is *comfortable* to read the page? Change the wattage of the light bulb. What effect does it have on the height you must raise it before you can read the page?

2. Obtain a light meter and measure the illumination levels in various offices, hallways, etc. How do they compare to the recommended levels given in Table 14-3, page 408 of the text?

3. Observe the use of VDTs in various offices. Measure the illumination levels and observe the screens for reflections. If reflections are present, how could they be reduced or eliminated?

Project 14

## DETERMINING ILLUMINATION LEVELS

### Reading Assignment

Text: Chapter 14, pages 394–411.

### Purpose

The purpose of this project is to illustrate the various concepts introduced in the text to a practical illumination problem.

### Problem

A hypothetical factory is expanding its operation. As part of this expansion an inspection station is to be moved to a new area of the plant. The type of activities which will be performed in this area include reading poorly reproduced orders, difficult, but not highly difficult, and inspection tasks involving inspection of small parts. Speed and accuracy are critical. The inspectors are between 40 and 55 years of age. The reflectance of the task background is only 20 percent.

*Part A.* Using Table 14-3, page 408 of the text, decide which category this type of activity represents.

Category _____

*Part B.* Using Table 14-4, page 409 of the text, determine the appropriate weighting factors of this situation.

Age _____

Speed/accuracy _____

Reflectance _____

Total weighting factor = _____

Using the appropriate rule on page 409 of the text, determine the recommended level of illuminance required in this area.

Recommended illuminance _____ 1x

*Part C.* One of the tasks performed by the inspectors is reading a micrometer. Since speed is critical, the plant manager wonders whether doubling the illuminance level over what you recommended in Part B would reduce the time required to read the micrometer. Based on Figure 14-10, page 406 of the text, what would you predict?

*Circle one:* Would/Would not    improve performance.

*Part D.* Using the formula on page 396 of the text for reflectance, determine the background luminance of the task.

Luminance = _____ cd/m²

*Part E.* The ceiling of the plant is 6 meters above the task surface. Using the formula for illuminance on page 395 of the text, determine the luminous intensity (candlepower) of a light source located at the ceiling that would be required to supply the level of illumination you recommended.

Candlepower = _____ cd

*Part F.* How much candlepower would be required if the light source was hung from the ceiling so that it was 3 meters above the task?

Candlepower = _____ cd

*Part G.* If the manager was planning to use hanging fluorescent lights, as described in Part F, as sources, how many watts of power would be required to illuminate this task? Use Figure 14-7 on page 399 of the text to select an average efficacy value for fluorescent lamps and use the following formula to compute the power requirements.

$$\text{Power required} = \frac{\text{Candlepower} \times 12.57}{\text{Lamp efficacy}}$$

Power required _____ W

*Part H.* If the plant keeps the lights on for 200 hrs per month, how many kilowatt-hr per month will be used to light this task? Use the following formula.

$$\text{Kilowatt-hr} = \frac{p \times t}{1000}$$

Where:
 p = power usage in watts
 t = time in hours

Kilowatt-hrs = _____

# Chapter 15

# ATMOSPHERIC CONDITIONS

**Terms/Concepts:**

Core temperature
Convection
Evaporation
Radiation
Conduction
Zone of cooling
Comfort zone
Zone of heat regulation
Acclimatization
Effective temperature
Operative temperature
Oxford index

Wet-bulb temperature
Wet-bulb globe temperature
Botsball
Heat stress index
Heat index
Wind chill index
Clo unit
Prescriptive zone
Hypoxia
Decompression sickness
Air ions

**Review Questions:**

1. Make a chart showing the various composite indices of heat stress and the factors that are considered by each.

2. Discuss the physiological and performance effects of heat stress.

3. Summarize the physiological and performance effects of cold stress.

4. Discuss some methods for reducing the negative consequences of cold stress.

5. Discuss the effects of high-altitude and under-sea atmospheric pressure on humans.

6. What are the alleged effects of air ions and what does the research suggest?

**Activities:**

1. Compare your performance at tying knots with string and picking up thin objects (such as a dime, paper clips, or pins) when your hands are very cold and when they are warm. Try it with your eyes closed.

2. Compute the WBGT for representative summer days in the city where you live. (You can determine wet-bulb temperature from dry-bulb temperature and relative humidity using Figure 15-3, page 432 of the text.) Compare the WBGTs to those recommended for continuous work shown in Figure 15-10, page 442 of the text.

Project 15

## EFFECTS OF TEMPERATURE ON MAN

### Reading Assignment

Text: Chapter 15, pages 427–444.

### Purpose

The purpose of this project is to familiarize the student with various indexes of environmental factors that affect the heat exchange process, and to acquaint the student with the kinds of predictions that can be made about human responses to various environmental conditions.

### Problem

A nationwide construction company has won three contracts to build buildings. Each of these contracts is in a different part of the country but the buildings will be built during the summer months. The company is concerned about the possible effects of the climatic conditions on the health and efficiency of the men. The company has obtained the information in Table 1 about the weather conditions in the building localities.

*Part A.* The information in Table 1 is not complete. Using Figure 15-3 on page 432 of the text, determine the values for the missing data and record them in Table 1.

*Part B.* Using Table 2 to record your answers, answer the following questions for each of the three locations given in Table 1. Note, the original effective temperature for each location has been supplied in Table 2 since the text does not contain a nomograph for determining it.

1. Using Figure 15-3 on page 432 of the text, determine the New Effective Temperature (ET*) for each location.

2. Using the formula given on page 431 of the text, determine the Oxford Index for each location.

3. Using the formula given on page 433 and the $t_g$ values given in Table 1, compute the Wet-Bulb Globe Temperature for each location. Assume that $t_{nwb}$ is equivalent to the wet-bulb temperature reported in Table 1.

4. Using Figure 15-4 on page 434 of the text and using the dry-bulb temperature as air temperature, determine the heat index in °F for each of the three locations.

5. Using the table on page 434 of the text, indicate the heat index category for each of the three locations.

6. Using Figure 15-6 on page 439 of the text, determine the mean rectal temperature which might be expected of the men if their work load is approximately 300 kcal/hr.

7. Using Figure 15-7 on page 439 of the text and dry-bulb temperature, estimate the mean heart rate of white and black males working in each of the three locations.

8. Based on Figure 15-9 on page 440 of the text, would any of the performances listed in Figure 15-9 be impaired after 80 minutes in each of the three locations.

9. Using Figure 15-10 on page 442 of the text and assuming moderate to heavy continuous work of 350 kcal/hr, determine for each location whether the recommended limit is exceeded.

*Part C.* The company is considering compensating the workers according to the physical stress incurred due to differences in climatic conditions of the building locations. Would you recommend that workers be paid differentially in the three locations? *If so, how and why. If not, why not.*

_____

_____

_____

**TABLE 1**
Basic Data on Weather Conditions for the Three Building Sites

| | Location | | |
| | A | B | C |
|---|---|---|---|
| Relative humidity | ———— | 30% | 80% |
| Wet bulb temperature (°F) | 75 | 66 | ———— |
| Dry bulb temperature (°F) | 90 | ———— | 69 |
| Globe temperature ($t_g$) | 97 | 91 | 78 |

**TABLE 2**
Summary of Information of Effects of Weather in the Three Building Locations

| | Location | | |
| | A | B | C |
|---|---|---|---|
| Original Effective Temperature (ET) | 81 | 77 | 67 |
| 1. New Effective Temperature (ET*) | ———— | ———— | ———— |
| 2. Oxford Index | ———— | ———— | ———— |
| 3. Wet-Bulb Glob Temperature (WBGT) | ———— | ———— | ———— |
| 4. Heat Index (°F) | ———— | ———— | ———— |
| 5. Heat Index category | ———— | ———— | ———— |
| 6. Mean rectal temperature (°F) | ———— | ———— | ———— |
| 7. Mean heart rate—white males | ———— | ———— | ———— |
| Mean heart rate—black males | ———— | ———— | ———— |
| 8. Performance impaired (Yes-No) | ———— | ———— | ———— |
| 9. Recommended limit exceeded (Yes-No) | ———— | ———— | ———— |

# Chapter 16

# NOISE

**Terms/Concepts:**

Noise

A sound-pressure scale

B sound-pressure scale

C sound-pressure scale

Phon

Sone

Perceived level of noise

Mark VII sone

Perceived noise level

Noy

Equivalent sound level

Nerve deafness

Conductive deafness

Audiometer

Presbycusis

Sociocusis

$TTS_2$

Effective quiet

PTS or NIPTS

Startle response

Funneling of attention

Gaps in performance

Noise dose

Partial dose

8-h time-weighted average sound level

Action level

Permissible exposure level

Impulse noise

Infrasonic noise

Ultrasonic noise

Day-night level

Normalized day-night level

Noise reduction rating

**Review Questions:**

1. Discuss the various psychophysical indices of loudness and distinguish between absolute and relative measures.

2. Describe the typical pattern of hearing loss that occurs with years of exposure to occupational noise. Indicate the relationship between temporary and permanent threshold shifts.

3. Summarize the effects of noise on performance. What specific effects have been found and what theories have been proposed to account for the effects?

4. Discuss the OSHA noise exposure criteria including the concepts of noise dose, TWA, action level and permissible exposure level.

5. Why is it so difficult to predict community response to annoying noise?

6. What are some techniques for controlling noise at the source, along its path, or at the receiver? How does the frequency of the noise effect these techniques?

**Activities:**

1. Spend a few minutes in a noisy environment, such as where circular saws or large motors are operating. Leave the area and see if you can detect a temporary threshold shift in hearing.

2. Obtain a sound-pressure meter and measure the sound level of some common environments, such as inside your house or apartment, along side a busy highway, or in the halls at school as classes let out. Measure the levels using the A, B, and C scales. How do they differ and why?

3. Obtain a frequency oscillator and place it a few feet in front of you. Select a high frequency (4000–5000 Hz) tone and listen to the intensity of it (set it at 80 dB). Place a piece of plywood (4×4 ft) between the oscillator and your face. Did you notice a reduction in the loudness of the tone? Try it again with a low frequency tone (100–300 Hz). What did you observe and why?

Project 16-A

## MEASUREMENT OF SOUND LOUDNESS AND NOISE DOSE

### Reading Assignment

Text: Chapter 16, pages 456–461, 470–471.

### Purpose

The purpose of this project is to illustrate the meaning and interrelationship of the sone and phon, and to demonstrate the concept of noise dose.

### Problem 1

A pure tone of 100 Hz has an intensity of 50 dB.

*Part A*. Using Figure 16-2, page 458 of the text, determine the loudness of the tone in phons.

Loudness = _____ phons

*Part B*. What intensity must a 500 Hz tone have to be of equal loudness?

Intensity = _____ dB

*Part C*. What is the relative loudness of the 100 Hz and 500 Hz tones in sones? (Consider the definition of a sone.)

Relative loudness = _____ sones

*Part D*. Is a sound of 80 phons twice as loud as a sound of 40 phons?

_____

*Part E*. How many phons would be twice as loud as 40 phons?

_____ phons

*Part F*. Look at Figure 16-2, page 458 of the text. Why do you think there is a difference between the data obtained from 20 years old and that obtained from 60 year olds at frequencies above 2000 Hz?

_____

_____

_____

## Problem 2

Two workers in a factory had exposures to noise over an 8 hr day as given in Table 1.

*Part A:* Using the formula and procedure discussed on pages 470–471, compute the noise dose for each worker.

Instructions:

1. Using Table 16-3, page 470 of the text, and the formula on page 470 determine the partial noise dose for each level of noise exposure and record them in Table 1.
2. Compute the noise dose for each worker, sum and multiply by 100 the partial noise doses for each worker and record in Table 1.

*Part B:* Using Table 16-4, page 471 of the text, convert the noise dose for each worker to the corresponding time-weighted average (TWA) and recode them in Table 1.

*Part C:* Which worker(s) exceed the action limit for noise exposure?

_____

*Part D:* Which worker(s) exceed the permissible exposure limit for noise exposure?

_____

### TABLE 1
#### Noise Exposure Data

**WORKER: FRANK W.**

| Sound Level, dBA | Exposure Time, hr | Partial Noise Dose |
|---|---|---|
| 75 | 1.0 | _____ |
| 80 | 0.9 | _____ |
| 85 | 1.6 | _____ |
| 90 | 4.0 | _____ |
| 95 | 0.4 | _____ |
| 100 | 0.1 | _____ |
| 105 | 0 | _____ |
| Noise Dose (sum of partial noise doses × 100) = | | _____ |
| TWA, dBA = | | _____ |

**WORKER: ROBERT L.**

| Sound Level, dBA | Exposure Time, hr | Partial Noise Dose |
|---|---|---|
| 75 | 3.0 | _____ |
| 80 | 0.5 | _____ |
| 85 | 0 | _____ |
| 90 | 0 | _____ |
| 95 | 2.0 | _____ |
| 100 | 2.0 | _____ |
| 105 | 0.5 | _____ |
| Noise Dose (sum of partial noise doses × 100) = | | _____ |
| TWA, dBA = | | _____ |

Project 16-B

## DAY-NIGHT LEVEL OF NOISE EXPOSURE AND ANNOYANCE

### Reading Assignment

Text: Chapter 16, pages 460–461, 473–477.

### Purpose

The purpose of this project is to demonstrate the calculations of the day-night level of noise exposure and the interpretation of the resulting annoyance.

### Problem

A corporation wishes to expand, and build a "widget" factory in a city. The company already owns a similar factory in another state and knows that the noise produced by such a factory might engender unfavorable reactions from residents in the area. The problem is to determine a suitable location for the factory. The company officials feel a responsibility to the residents living in the area around the factory as well as to the employees who will work in the factory. A trade-off exists. On the one hand the company must consider the distance workers must travel to get to work, as well as the problems of traffic and parking for them. The further from suburban areas the factory is located, the greater the distance the workers would have to travel, the greater the traffic problem and the greater the parking problem.

On the other hand, the closer to the suburban areas the factory is located, the greater the chance that people will complain and bring legal action against the factory because of the noise.

From noise surveys of the other factory the company predicts that the new factory will have a day-night level of noise equal to 68 dBA.

*Part A.* Using Table 16-7, page 476 of the text, determine the normalized day-night values for the various combinations of correction factors shown in Table 2. Make no correction for season as the plant will operate in both summer and winter. Make no correction for pure tone or impulse noise. Record the normalized values in the upper right corner of each square in Table 2.

*Part B.* For each normalized day-night level in Table 2, determine the expected community response as given in Table 16-8, page 476 of the text. Record them in the boxes of Table 2.

*Part C.* Assume that the company does not want to provoke any threats of community action against the company initially, and further, they do not want widespread complaints even after considerable exposure to the noise. They feel that sporadic complaints from residents can be alleviated by engaging in various civic improvement activities, such as donating to and participating in, community acitivites. Based on these considerations, in what location would you recommend the company locate its factory?

Recommended location ————————————

**TABLE 2**
Normalized Day-Night Levels

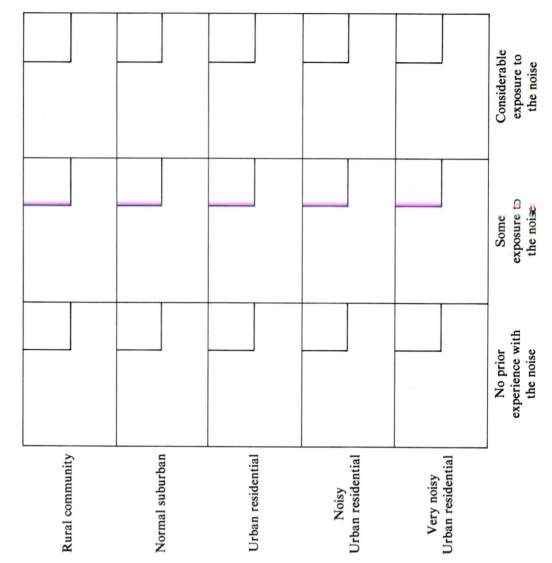

Rural community

Normal suburban

Urban residential

Noisy
Urban residential

Very noisy
Urban residential

No prior
experience with
the noise

Some
exposure to
the noise

Considerable
exposure to
the noise

Normalized L_{dn}

Community response*

*Key:

No reaction or
sporadic complaint—None

Widespread
complaints—Widespread

Severe threats—Severe

Vigorous action—Vigorous

# Chapter 17

# MOTION

**Terms/Concepts:**

Exteroceptors

Proprioceptors

Kinesthetic receptors

Semicircular canals

Vestibular sacs (or otolith organs)

Utricle

Saccule

Sinusoidal vibration

Random vibration

$g_x$

$g_y$

$g_z$

Power spectral density

Root-mean-square acceleration

Resonance

Amplification

Attenuation

Equal-comfort contours

DISC units

Fatigue-decreased proficiency

Acceleration

Linear acceleration

Rotational acceleration

Radial acceleration

Angular acceleration

$G_x$

$G_y$

$G_z$

Anti-G suit

Second collision

Space adaptation syndrome

Disorientation

Coriolis illusion or cross-coupling effect

Oculogravic illusion

Autokinesis

Head symptoms

Sopite syndrome

Gut symptoms

Sensory rearrangement theory

**Review Questions:**

1. Discuss the physiological and performance effects of vibration.
2. Discuss the general effects of acceleration in each of the three directions.
3. List some techniques for reducing the negative effects of acceleration.
4. Describe the important features of the typical relaxed posture assumed in a weightless environment.
5. Discuss some examples of illusions caused by motion.
6. What causes motion sickness?

**Activities:**

1. Ride a roller coaster and when it accelerates down a steep hill try to execute some of the movements pictured in Figure 17-15, page 503 of the text. Which movements seem harder to make? Why?
2. When you next have a chance to take a boat ride in rolling water, determine the number of cycles (up and down) the boat goes through per minute. Compare this with Figure 17-21, page 513 of the text. (Note Figure 17-21 is in cycles per second, so divide cycles per minute by 60

to convert to cycles per second.) Stand in the middle of the boat and at the bow or stern. Did you feel better near the middle? Consider the differences in amplitude of the boat motion at the various locations on the boat. Describe any symptoms of motion sickness you experienced. What seemed to reduce the symptoms?

3. Next time you see pictures of astronauts in space, compare their posture to that shown in Figure 17-20, page 509 of the text.

## Project 17

## ASSESSING RIDE DISCOMFORT

### Reading Assignment

Text: Chapter 17, pages 494–496.

### Purpose

The purpose of this project is to more fully illustrate the technique for determining ride discomfort introduced in the text.

### Problem

A bus company has a chance to buy a new bus but they are not sure which of two models (A or B) to buy. They are concerned that the least expensive model, B, costing 50 percent less, may cause more discomfort. However, if the increase in discomfort is not too much, then it may be more cost effective to purchase the less expensive bus. They have received the vibration data shown in Table 1 for the two buses.

*Part A.* Using the vertical and lateral vibration data in Table 1, compare the buses to the ranges for buses depicted in Figure 17-3, page 490 of the text. Would you say they are below average, average, or above average in terms of intensity of vibration?

Bus A _____

Bus B _____

*Part B.* For each bus, compute the ride discomfort in DISC units for a 5 and 60 minute ride. To make this a little easier, we will assume that the vibration is sinusoidal in all three axis.

Procedure:[1]

1. Convert the rms g accelerations for vertical and lateral vibration to peak g values. The roll vibration is not converted. Use the formula in the footnote on page 489 of the text. Enter the values in Table 1.
2. Compute the DISC values for each axis separately using the following formulas. (Note, these formulas are specific to the frequencies and accelerations shown in Table 1.) Record the DISC values in Table 1.

   $DISC_{vert} = 10.41 \times peak\ g$
   $DISC_{lat} = 8.12 \times peak\ g$
   $DISC_{roll} = -0.18 + (4.7 \times acceleration)$

3. Combine the single axis DISC values into a total DISC using the following formula. Record the results in Table 2.

   $DISC_{total} = 1.14 \times DISC_{vlr}$

   where:

   $DISC_{vlr} = \sqrt{DISC^2_{vert} + DISC^2_{lat} + DISC^2_{roll}}$

1. From Leatherwood, J., Dempsey, T., and Clevenson, S. A design tool for estimating passenger ride discomfort within complex ride environments. *Human Factors* 1980, *22*, 291–312.

4. Compute the ride discomfort after 5 minutes and after 60 minutes. First compute a correction factor ($\Delta DISC_{dur}$) for ride duration using the following formula:

$$\Delta DISC_{dur} = 0.003 - 0.012t$$

where: t = duration of ride in minutes

Compute the corrected DISC values using the following formula:

$$DISC_{corr} = DISC_{total} + \Delta DISC_{dur}$$

Do this for both buses using a 5 min and 60 min ride duration. Record the values in Table 2.

*Part C.* Using Table 17-2, page 494 of the text, determine for each bus the percentage of passengers that would feel uncomfortable after 5 min and after 60 min.

Bus A: after 5 min _____%; after 60 min _____%

Bus B: after 5 min _____%; after 60 min _____%

*Part D.* How many times more uncomfortable is Bus B than Bus A after 5 min and after 60 min?

After 5 min _____ times more uncomfortable

After 60 min _____ times more uncomfortable

Is there any question now as to which bus to buy, even with the 50 percent cost saving with Bus B?

### TABLE 1
#### Basic Vibration Data on Two Buses.

| Axis of Vibration | Frequency Hz | Acceleration | Peak g | DISC |
|---|---|---|---|---|
| Bus A: | | | | |
|     Vertical | 4 | 0.05 rms g | _____ | _____ |
|     Lateral | 6 | 0.05 rms g | _____ | _____ |
|     Roll | 2 | 0.10 rad/s² | | _____ |
| Bus B: | | | | |
|     Vertical | 4 | 0.08 rms g | _____ | _____ |
|     Lateral | 6 | 0.16 rms g | _____ | _____ |
|     Roll | 2 | 0.10 rad/s² | | _____ |

### TABLE 2
#### Ride Discomfort Values

| | $DISC_{total}$ | $DISC_{5min}$ | $DISC_{60min}$ |
|---|---|---|---|
| Bus A | _____ | _____ | _____ |
| Bus B | _____ | _____ | _____ |

# Chapter 18

# HUMAN FACTORS APPLICATIONS
# IN SYSTEM DESIGN

**Terms/Concepts:**

Design process stages
Functional flow diagrams
Allocation of functions
Affective support
Cognitive support
Task analysis
Operational-sequence diagram
CAFES
FAM
WAM
Engineering standards
MIL-SPECS
Practical significance
IRA
CAR

Embedded training
Pictogram
Evaluation
Controlled experimentation
Personnel subsystem measurement
Products liability
"But for" test
Patent-danger rule
Unreasonable danger
Warnings
Instructions
Levels of hazard
Fundamental elements of a warning
Warning overload

**Review Questions:**

1. What are the 6 major stages in system design and how do they relate to one another?

2. Discuss the allocation of function process and the utility of "relative capability lists."

3. What are some problems and considerations involved in applying human factors data to system design?

4. What is the legal definition of a defective product and what implications does this have for human factors and the design of products?

5. What are some considerations that need to be taken into account when designing warnings on consumer products?

**Activities:**

1. Look at warnings contained on various consumer products. Evaluate them with respect to the considerations discussed on pages 547–552 of the text.

2. Keep a look-out for reports of products liability cases in the newspaper. Consider the case in terms of the discussion on pages 543–546 of the text.

Project 18

## WRITING WARNINGS

### Reading Assignment

Text: Chapter 18, pages 547–552.

### Purpose

The purpose of this project is to provide opportunities to critique and rewrite various warnings based on the principles presented in the text.

### Problem

*Part A.* The following warning was placed on a bottle of toxic, flammable liquid. Evaluate this warning based on the requirements presented in the text.

| CAUTION: Handle with care |
| --- |

_____

_____

_____

_____

_____

*Part B.* An air mixture ring on a gas burner is to be rotated to adjust the temperature of the flame. Rotating the ring changes the size of the holes which controls the amount of gas and air that mix together. The larger the holes, the hotter the flame. The person adjusting the ring needs to be careful because the ring becomes hot and takes quite a while to cool down after the flame is extinguished. Write a warning which would appear on the gas burner advising the user of this situation.

*Part C.* Each year many people are crippled when they run and dive into the surf at the beach. Hidden sand bars, inshore holes, and other irregular bottom conditions are struck by the person diving and the force results in a broken neck. Most end up as quadaplegics. Write a warning advising beach goers of this situation.

*Part D.* The following warning was displayed on a machine because if undue force was used to tighten the bolts, the bolts could break and damage the equipment. Rewrite the warning message.

> Caution: It is important that undue force not be used when bolts are tightened.

*Part E.* The following is an actual warning from a bottle of correction fluid ("white-out"). Rewrite the warning.

> WARNING; Contains Trichoroethane. Intentional misuse by deliberately concentrating, sniffing or inhaling contents can be harmful or fatal. Keep out of reach of children.

# Chapter 19

# THE BUILT ENVIRONMENT

**Terms/Concepts:**

Open-plan office
Bull pen office
Landscaped office

Persons per room (PPR)
Square feet per person (SFPP)

**Review Questions:**

1. What are some functions and tasks performed by office workers and what implications do they have for design of the office environment?
2. What are some advantages and limitations of landscaped and bull pen offices?
3. What are some considerations in the design and layout of dwelling units?

**Activities:**

1. Visit office buildings that employ open office plans and observe the behavior and working environment. Talk to some of the workers about their perceptions of the office and limitations and advantages of the particular design.
2. Measure the dimensions of various inside and outside stairways and compare them to the recommendations given on page 573 of the text.
3. Visit a retirement home and discuss with the staff and residents what special design considerations are involved in housing for the elderly and handicapped.

Project 19

## A PRIVATE DWELLING UNIT

### Reading Assignment

Text: Chapter 19, pages 566–573.

### Purpose

The purpose of this project is to provide an opportunity to critically evaluate and apply human factors concepts discussed in the text to an actual private dwelling floor plan.

### Problem

Below is an actual floor plan[1] for a residential single family house with no basement or attic. It has approximately 2,600 square feet of living space. A house like this is usually sold to a couple with at least one small child. Such families usually do a lot of entertaining in their homes, including giving moderate sized dinner parties. Rarely would a family who buys such a home have live-in help.

Critically evaluate this floor plan, noting both desirable and not-so desirable features. (Use additional paper if you wish.) Consider *among other things,* probable traffic flow, convenience of facilities, closet and storage space, facilities and traffic flow for such activities as entertaining, child care, and nighttime emergencies, and overall use of available space.

1. From Metropolitan Development Corporation, Brightwood Development, 1976. Reprinted by permission.

# Chapter 20

# HIGHWAY TRANSPORTATION
# AND RELATED FACILITIES

**Terms/Concepts:**

Accident

Accident proneness

Field dependence—field independence

Adaptation

System geometry of roadway illumination

Cutoff distance

Visibility distance

Dwell time

**Review Questions:**

1. What are the major human causes of automobile accidents?

2. What are some of the human variables that seem to be related to driving performance?

3. List examples of human factors concerns in the design of urban transit systems?

**Activities:**

1. Be a passenger in a car driven by a friend. Have the friend drive at 55 mph for at least 10 minutes and then slow down to some speed between 20 and 50 mph. How close can you guess the speed? Did you over or under estimate it? Drive at 30 mph for about 10 minutes and then have your friend speed up. Try to guess the speed again. Did your responses illustrate the phenomenon of adaptation as described in the text?

2. Go to a bus station, train station, subway, or airport and evaluate the system from a human factors perspective. Evaluate it in terms of the station or terminal facilities and features, and the interior of the vehicles. Pay particular attention to the needs of the handicapped.

Project 20

# PREFERENCES FOR PUBLIC TRANSPORTATION SYSTEM CHARACTERISTICS

## Reading Assignment

Text: Chapter 20, pages 598–601.

## Purpose

The purpose of this project is to illustrate the relative importance of various public transportation system characteristics by collecting opinion data using the paired comparison technique.

## Problem

A city transit authority is interested in updating the preference data of Golob, Canty, and Gustafson (1970) reported in the text on page 599. Further, to overcome the difficulty of combining scale values across blocks of items, as discussed on page 599 of the text, fewer items will be evaluated and all will be evaluated in the same set.

The items that were of most concern to the transit authority are listed in Table 1.

*Part A.* Using the paired comparison technique, determine scale values for each of the 13 items in Table 1.

Procedure:

1. Cut out the paired comparison items on the last two pages of this project. Each one is a 2 in × 1 in rectangle with two characteristics listed on it. There are 78 items in all, representing all possible pairs of characteristics.
2. Mix up the items and place them face down in a pile on a table.
3. Use a friend as the subject.
4. Read the person the following instructions:

   > Listed on each of these slips of paper are two characteristics of public transportation systems. Turn over one slip at a time and choose which of the two characteristics *you* would prefer to have in a public transit system. Read the number and item that you choose. You must choose one; there can be no ties. Do this for each of the slips. Are there any questions?

5. Record the person's choices in column 3 of Table 1 ( "count"). Make a "tick" mark each time an item is chosen.
6. Tally the tick marks for each item and record the totals in column 4 of Table 1. (The maximum count for any item is 12.)
7. The total number of tick marks for an item corresponds to its scale value. The more tick marks, the higher the preference for the item. (A more complex procedure can be used to compute scale values, however, for our purposes, the effort is not worth the increase in information gained.)
8. This procedure can be repeated with several subjects and the average number of tick marks for each item can be computed and used in the next section of this project.

*Part B.* Compare the scale values you obtained with those reported in Figure 20-10, page 599 of the text.

Procedure:

1. Rank the scale values you obtained using a rank of 1 for the most preferred. In cases of tied scale values, give each item the average of the ranks that would have been assigned to the items if they were not tied. For example, if two items are tied for 6th place, then assign both a rank of 6.5, i.e., $(6+7)/2$. Record the ranks in column 5 of Table 1.
2. Find the 13 items used in this project in Figure 20-10, page 599 of the text. Record their scale values in column 6 of Table 1.
3. Rank the scale values in column 6 and record the ranks in column 7.
4. Compare the ranks you obtained and those from the text.

What are the similarities and differences? What might account for the differences you found?

_____

_____

_____

_____

_____

_____

_____

_____

_____

## TABLE 1
### Worksheet for Recording Preferences

| (1)<br>Item | (2)<br>Description | (3)<br>Count | (4)<br>Total | (5)<br>Rank | (6)<br>Text Values | (7)<br>Rank |
|---|---|---|---|---|---|---|
| 1 | Arriving when planned | | | | | |
| 2 | No transfer trip | | | | | |
| 3 | Lower fares | | | | | |
| 4 | Shorter walk to pick up | | | | | |
| 5 | Having a seat | | | | | |
| 6 | Shelters at pick up | | | | | |
| 7 | Easy fare paying | | | | | |
| 8 | Coffee, newspaper, etc. on board | | | | | |
| 9 | Easy entry / exit | | | | | |
| 10 | Space for packages | | | | | |
| 11 | Adjustable seats | | | | | |
| 12 | Form talking groups while riding | | | | | |
| 13 | Stylish vehicle exterior | | | | | |

|  |  |  |  |  |
|---|---|---|---|---|
| 1—Arriving when planned<br>2—No transfer trip | 3—Lower fares<br>1—Arriving when planned | 1—Arriving when planned<br>4—Shorter walk to pick up | 5—Having a seat<br>1—Arriving when planned | 1—Arriving when planned<br>6—Shelters at pick up |
| 7—Easy fare paying<br>1—Arriving when planned | 1—Arriving when planned<br>8—Coffee, newspaper, etc. on board | 9—Easy entry/exit<br>1—Arriving when planned | 1—Arriving when planned<br>10—Space for packages | 11—Adjustable seats<br>1—Arriving when planned |
| 12—Form talking groups while riding<br>1—Arriving when planned | 1—Arriving when planned<br>13—Stylish vehicle exterior | 2—No transfer trip<br>3—Lower fares | 4—Shorter walk to pick up<br>2—No transfer trip | 2—No transfer trip<br>5—Having a seat |
| 6—Shelters at pick up<br>2—No transfer trip | 2—No transfer trip<br>7—Easy fare paying | 8—Coffee, Newspaper, etc. on board<br>2—No transfer trip | 2—No transfer trip<br>9—Easy entry/exit | 10—Space for packages<br>2—No transfer trip |
| 2—No transfer trip<br>11—Adjustable seats | 12—Form talking groups while riding<br>2—No transfer trip | 2—No transfer trip<br>13—Stylish vehicle exterior | 3—Lower fares<br>4—Shorter walk to pick up | 5—Having a seat<br>3—Lower fares |
| 3—Lower fares<br>6—Shelters at pick up | 7—Easy fare paying<br>3—Lower fares | 3—Lower fares<br>8—Coffee, newspapers, etc. on board | 9—Easy entry/exit<br>3—Lower fares | 3—Lower fares<br>10—Space for packages |
| 11—Adjustable seats<br>3—Lower fares | 3—Lower fares<br>12—Form talking groups while riding | 13—Stylish vehicle exterior<br>3—Lower fares | 4—Shorter walk to pick up<br>5—Having a seat | 6—Shelters at pick up<br>4—Shorter walk to pick up |
| 4—Shorter walk to pick up<br>7—Easy fare paying | 8—Coffee, newspaper, etc. board<br>4—Shorter walk to pick up | 4—Shorter walk to pick up<br>9—Easy entry/exit | 10—Space for packages<br>4—Shorter walk to pick up | 4—Shorter walk to pick up<br>11—Adjustable seats |
| 12—Form talking groups while riding<br>4—Shorter walk to pick up | 4—Shorter walk to pick up<br>13—Stylish vehicle exterior | 5—Having a seat<br>6—Shelters at pick up | 7—Easy fare paying<br>5—Having a seat | 5—Having a seat<br>8—Coffee, newspaper, etc. on board |

| | | | | | | | | |
|---|---|---|---|---|---|---|---|---|
| 13—Stylish vehicle exterior<br>5—Having a seat | 6—Shelters at pick up<br>11—Adjustable seats | 7—Easy fare paying<br>10—Space for packages | 10—Space for packages<br>8—Coffee, newspaper, etc. on board | 11—Adjustable seats<br>9—Easy entry/exit | | | | |
| 5—Having a seat<br>12—Form talking groups while riding | 10—Space for packages<br>6—Shelters at pick up | 9—Easy entry/exit<br>7—Easy fare paying | 8—Coffee, newspaper, etc. on board<br>9—Easy entry/exit | 9—Easy entry/exit<br>10—Space for packages | | | | |
| 11—Adjustable seats<br>5—Having a seat | 6—Shelters at pick up<br>9—Easy entry/exit | 7—Easy fare paying<br>8—Coffee, newspaper, etc. on board | 13—Stylish vehicle exterior<br>7—Easy fare paying | 8—Coffee, newspaper, etc. on board<br>13—Stylish vehicle exterior | 10—Space for packages<br>11—Adjustable seats | 11—Adjustable seats<br>12—Form talking groups while riding | | |
| 9—Easy entry/exit<br>5—Having a seat | 6—Shelters at pick up<br>7—Easy fare paying | 12—Form talking groups while riding<br>6—Shelters at pick up | 11—Adjustable seats<br>7—Easy fare paying | 8—Coffee, newspaper, etc. on board<br>11—Adjustable seats | 9—Easy entry/exit<br>12—Form talking groups while riding | 12—Form talking groups while riding<br>10—Space for packages | 13—Stylish vehicle exterior<br>11—Adjustable seats | |

# Chapter 21

# HUMAN ERROR AND WORK-RELATED TOPICS

**Terms/Concepts:**

Human error
Errors of omission
Errors of commission
Sequence error
Timing error
Exclusion designs
Prevention designs
Fail-safe designs
Signal

False alarms
Limit samples
Critical incident technique
Fault-tree analysis
Feedback
Contingency reinforcement
Incentives
Propaganda

**Review Questions:**

1. Discuss various human-error classification schemes. What are their strengths and weaknesses?

2. Given the research findings reviewed in the text, how would you design an inspection activity for maximum performance?

3. Discuss the relationships between training, feedback, reinforcement, incentives, and propaganda in reducing accidents.

**Activities:**

1. Briefly describe an accident that took place in the home or automobile or at work. Classify it according to the various schemes presented in the text, and analyze its cause using the model presented in Figure 21-8, page 622 of the text, and the breakdown used to analyze accidents in South African gold mining and presented on page 623 of the text.

Project 21-A

## CLASSIFICATION OF ERRORS

### Reading Assignment

Text: Chapter 21, pages 607–610.

### Purpose

The purpose of this project is to familiarize the student with one system of classifying errors discussed in the text and to illustrate the subjective evaluation required to use the system.

### Problem

In the system developed by Rook (see text, page 608) errors are classified in a two-dimensional matrix. The classification of a particular error into a specific cell of the matrix is not absolute. Some subjective evaluation is involved. In some cases the same objective error can be classified into different cells depending on other factors involved in the case.

*Part A.* Listed below are 13 short descriptions of hypothetical errors which can and do occur in various situations. For each one, decide how you would classify the error using the system discussed in the text (page 609). Table 1 reproduces the classification system as presented in the text. Enter the appropriate code for each item in the space provided. If you feel the error could be classified in more than one cell, place the appropriate code in the "second choice" column to the right of the item.

Each of the first three errors is classified for you with a brief explanation of why it should be classified as it is.

**TABLE 1**
Classification Scheme

| | Behavior Component | | |
| --- | --- | --- | --- |
| | Input | Mediation | Output |
| A. Intentional | AI | AM | AO |
| B. Unintentional | BI | BM | BO |
| C. Omission | CI | CM | CO |

|  | First Choice | Second Choice |
| --- | --- | --- |

1. The wrong resistor was used in assembling a radio. It was discovered that the color code on the resistor had faded and the colors were not very distinguishable.    First Choice: __AI__   Second Choice: _____

> This is an example of an *Intentional Input* error. The operator intended to use the resistor he chose, but it was the wrong one. Note: That Intentional—Unintentional refers to the act, not the consequences of the act. The error was at the input level, since he misread the color code. Note there is no attempt to fix the blame here, it could have been the operator's fault or the manufacturer's fault.

|  | First Choice | Second Choice |
|---|---|---|

2. A student during a test meant to subtract two numbers, but instead added them and hence got the wrong answer.    **BM** _____

> This is an example of an *Unintentional Mediation* error. The student knew what he wanted to do, but unintentionally did the wrong thing. The error was at the mediational level as it involved mental calculations.

3. An inspector measured the tolerance of a drill hole but forgot to write it down on the inspection report.    **CO** _____

> This is an example of an *Omission Output* error. The operator forgot to do something. The error was at the output level, since he failed to write it down.

4. An operator reversed two wires in an electrical connector. It was found that one wire was black and the other dark blue, almost impossible to tell apart. The operator knew which color went to which connector, but thought the blue wire was black.    _____ _____

5. A burned insulation on a wire was found. The cause was traced to the accidental contact of the soldering iron with the insulation of the wire.    _____ _____

6. A mail clerk missorted a letter addressed to South Dakota, placing it into the pigeon hole for North Dakota. The address on the envelope was handwritten and practically illegible. The clerk thought it said North Dakota.    _____ _____

7. The same error as in No. 6 was made except this time the address on the envelope was typewritten. The clerk knew it said South Dakota, but thought South Dakota's mail was to be place in pigeon hole 7—actually South Dakota's mail belonged in pigeon hole 8.    _____ _____

8. The same error as in No. 6 was made also with an envelope that was typewritten. In this case, however, the clerk knew which pigeon hole it belonged in, but when he tossed the letter someone bumped his arm and he missed the hole. It landed in the North Dakota pigeon hole instead of in the South Dakota pigeon hole.    _____ _____

9. An operator failed to set a switch during the checkout of a piece of equipment. It was found that there was a grease smudge on the checkout list which covered the instruction to throw the switch.    _____ _____

10. A press jammed when the speed control was advanced too rapidly. It was discovered that the bar had stuck, and the operator felt he could tap it loose with a hammer. The tap proved too great, causing the speed control to advance too rapidly.    _____ _____

11. An experienced operator forgot to solder in a transistor while assembling a radio.    _____ _____

138

138

|  | First Choice | Second Choice |
|---|---|---|

12. An expensive piece of equipment burned out from an electrical overload. _____ _____
The cause of the malfunction was traced to an operator who accidentally picked up the wrong operations manual. The manual he picked up was for a model 347R, but his machine was a model 870X. The manuals looked the same except for the number on the corner of the cover. The procedure outlined in the 347R manual, when applied to the 870X machine, caused an electrical overload.

13. An operator working on a radar scope reported an aircraft's bearing, but _____ _____
forgot to compute its speed. A near collision resulted.

*Part B.* Describe, in sufficient detail, a situation in which you made an error and classify it into the above system.

_____

_____

_____ Classification: _____

*Part C.* A major function of error classification systems is to direct efforts to reduce the likelihood of errors. List possible measures which might be taken to reduce the following errors described in Part A of this project.

1. Item 7 (forgot South Dakota pigeon hole number)

_____

_____

_____

2. Item 8 (bumped arm while throwing mail)

_____

_____

_____

3. Item 12 (picked up wrong manual)

_____

_____

_____

Project 21-B

## HUMAN FACTORS EVALUATION OF AN ELECTRONICS ASSEMBLY PLANT

### Reading Assignment

Text: Chapter 1-21, pages 1-633.

### Purpose

The purpose of this project is to provide an opportunity to integrate much of the material in the text to a single applied problem. Special emphasis is given to the material in Chapter 21.

### Problem

You receive a telephone call from the owner of a small electronics assembly plant in your city who wants you to help him with some problems he is having. During the short conversation you learn the following facts:

1. The plant is housed in a large converted warehouse.
2. Approximately 60 people are employed in the assembly plant area.
3. The basic flow of product assembly is as follows. Components are soldered on to circuit boards by workers sitting at long tables. The boards are then inspected using a magnifying glass for unacceptable solder joints and broken wires. In another part of the plant, the cabinets are spray painted and placed in an oven to bake. These are then inspected for paint defects such as bubbles, scratches, and too thin or thick areas of paint. The circuit boards and cabinets are then brought to workstations and assembled together using common hand tools. The completed assemblies are boxed and shipped on pallets.
4. The owner is having the following problems:
   a. Too many defective parts are being missed during inspection of the circuit boards and too many good cabinets are being rejected by the paint inspectors.
   b. Workers who solder components on the circuit boards and those who assemble them into the cabinets complain of wrist pain, aches in the shoulder, back, and neck.
   c. Several times the workers at the paint station have let the oven over heat, destroying the cabinets being baked.
   d. Workers are complaining about the noise generated by the two large fans used to help keep the paint area around the ovens cool.
   e. There have been several reported back injuries from lifting boxes of assembled parts onto pallets.
   f. There has been a rash of eye injuries because the workers do not wear their protective glasses and snipped pieces of wire hit them in the eye.

After relating his tale of woe, the owner asks that you come to the plant, look around, take measurements, and possibly even suggest some things which might improve the situation. To prepare for the visit, list for each of the owners problems (1) some things which might be causing the problem; (2) measurements and observations you want to make; and (3) some things to discuss with the owner which might improve the situation.

*Problem 1.* Too many defective boards being missed and too many good cabinets being rejected.

*Problem 2.* Aches and pains in the assembly workers.

*Problem 3*. Oven allowed to over heat.

*Problem 4*. Noise from the fans.

*Problem 5*. Back injuries from lifting boxes.

*Problem 6*. Eye injuries because safety glasses are not worn.